A Student's Approach to PreCalculus Student Solutions Manual

First Edition

A Student's Approach to PreCalculus Student Solutions Manual

Alfred K. Mulzet, Ph.D.

Palm Coast Publishing, Inc.
7701 Timberlin Park Blvd Unit 1212
Jacksonville, Florida 32256
USA

Copyright © 2018   Palm Coast Publishing, Incorporated
7701 Timberlin Park Blvd Unit 1212
Jacksonville, FL 32256

Printed in the United States of America
10 9 8 7 6 5 4 3 2 1

ISBN: 978-0-9776973-4-2

Palm Coast Publishing Incorporated, Jacksonville

Contents

Introduction

This book is intended to be a companion to A Student's Approach to Precalculus. It provides fully worked out solutions to select even number problems from the textbook. Students should be able to find a similar worked problem to every type of problem that appears in the text.

Chapter 1

Solving Equations and Inequalities

1.1 Exercises

In problems 1 - 30, solve for x.

2.
$$3x + 5 = 23$$
$$3x + 5 - 5 = 23 - 5$$
$$3x = 18$$
$$\frac{3x}{3} = \frac{18}{3}$$
$$x = 6$$

12.
$$5(x - 3) = 15$$
$$\frac{5(x - 3)}{5} = \frac{15}{5}$$
$$x - 3 = 3$$
$$x - 3 + 3 = 3 + 3$$
$$x = 6$$

20.
$$6(x - 3) - 4 = 6x + 15$$
$$6x - 18 - 4 = 6x + 15$$
$$6x - 22 = 6x + 15$$
$$6x - 22 + 22 = 6x + 15 + 22$$
$$6x = 6x + 37$$
$$6x - 6x = 6x + 37 - 6x$$
$$0 = 37$$

Since the variable dropped out of the equation in problem 20 and the resulting equation is false, the original equation has no solution.

In problems 22 and 30, we first multiply both sides of the equation by the least common denominator, then simplify both sides of the equation.

22.
$$\frac{1}{4}x - \frac{3}{4} = 2$$
$$4\left(\frac{1}{4}x - \frac{3}{4}\right) = 4 \cdot 2$$
$$x - 3 = 8$$
$$x - 3 + 3 = 8 + 11$$
$$x = 19$$

30.
$$\frac{3x+4}{5} = \frac{2}{3}x - 4$$
$$15\left(\frac{3x+4}{5}\right) = 15\left(\frac{2}{3}x - 4\right)$$
$$3(3x+4) = 10x - 60$$
$$9x + 12 = 10x - 60$$
$$9x + 12 - 12 = 10x - 60 - 12$$
$$9x = 10x - 72$$
$$9x - 10x = 10x - 72 - 10x$$
$$-x = -72$$
$$x = 72$$

In problems 31 - 60, solve the inequality. Then write the solution in interval notation, and graph it on a number line.

32.
$$\begin{aligned} 3x + 5 &< 23 \\ 3x &< 18 \\ \frac{3x}{3} &< \frac{18}{3} \\ x &< 3 \end{aligned}$$

Interval notation: $(-\infty, 3)$

Graph:

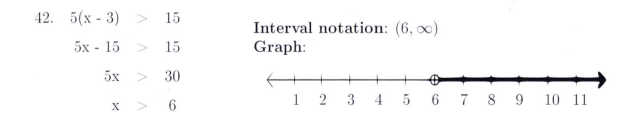

42.
$$\begin{aligned} 5(x - 3) &> 15 \\ 5x - 15 &> 15 \\ 5x &> 30 \\ x &> 6 \end{aligned}$$

Interval notation: $(6, \infty)$

Graph:

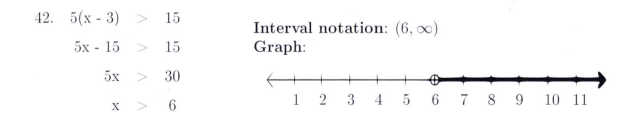

52.
$$\begin{aligned} \frac{1}{4}x - \frac{3}{4} &\leq 2 \\ 4\left(\frac{1}{4}x - \frac{3}{4}\right) &\leq 4 \cdot 2 \\ x - 3 &\leq 8 \\ x &\leq 11 \end{aligned}$$

Interval notation: $(-\infty, 11]$

Graph:

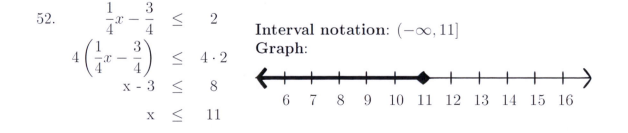

60.
$$\frac{3x+4}{5} \le \frac{2}{3}x - 4$$

$$15\left(\frac{3x+4}{5}\right) \le 15\left(\frac{2}{3}x - 4\right)$$

$$9x + 12 \le 10x - 60$$

$$9x \le 10x - 72$$

$$-x \le -72$$

$$x \ge 72$$

Interval notation: $[72, \infty)$

Graph:

1.2 Exercises

For exercises 1 - 30, find the solution set.

2. $|3x + 5| = 23$

$$3x + 5 = -23 \quad \text{or} \quad 3x + 5 = 23$$

$$3x = -28 \quad \text{or} \quad 3x = 18$$

$$x = -\frac{28}{3} \quad \text{or} \quad x = 6$$

Solution set : $\left\{\frac{28}{3}, 6\right\}$

12. $5|x - 3| = 15$

$$|x - 3| = 5$$

$$x - 3 = -5 \quad \text{or} \quad x - 3 = 5$$

$$x = -2 \quad \text{or} \quad x = 8$$

Solution set : $\{-2, 8\}$

22. $\left|\frac{1}{4}x - \frac{3}{4}\right| = 2$

$$\frac{1}{4}x - \frac{3}{4} = -2 \quad \text{or} \quad \frac{1}{4}x - \frac{3}{4} = 2$$

$$4\left(\frac{1}{4}x - \frac{3}{4}\right) = 4(-2) \quad \text{or} \quad 4\left(\frac{1}{4}x - \frac{3}{4}\right) = 4(2)$$

$$x - 3 = -8 \quad \text{or} \quad x - 3 = 8$$

$$x = -5 \quad \text{or} \quad x = 11$$

Solution set : $\{-5, 11\}$

30.

$$\left| \frac{3x+4}{5} \right| = \frac{2}{3}$$

$$\frac{3x+4}{5} = -\frac{2}{3} \quad \text{or} \quad \frac{3x+4}{5} = \frac{2}{3}$$

$$15\left(\frac{3x+4}{5} \right) = 15\left(-\frac{2}{3} \right) \quad \text{or} \quad 15\left(\frac{3x+4}{5} \right) = 15\left(\frac{2}{3} \right)$$

$$9x+12 = -10 \quad \text{or} \quad 9x+12 = 10$$

$$9x = -22 \quad \text{or} \quad 9x = -2$$

$$x = -\frac{22}{9} \quad \text{or} \quad x = -\frac{2}{9}$$

$$\textbf{Solution set}: \left\{ -229, -\frac{2}{9} \right\}$$

For exercises 31 - 60, solve for x and graph the solution set on a number line.

32.

$$|3x+5| < 23$$
$$-23 < 3x+5 < 23$$
$$-28 < 3x < 18$$
$$-\frac{28}{3} < x < 6$$

Interval notation: $\left(-\frac{28}{3}, 6 \right)$

Graph:

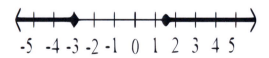

38.

$$|4x+3| \geq 9$$
$$4x+3 \leq -9 \quad \text{or} \quad 4x+3 \geq 9$$
$$4x \leq -12 \quad \text{or} \quad 4x \geq 6$$
$$x \leq -3 \quad \text{or} \quad x \geq \frac{3}{2}$$

Interval notation: $(-\infty, -3) \cup \left(\frac{3}{2}, \infty \right)$

Graph:

42.

$$15 < 5|x-3|$$
$$5 < |x-3|$$
$$|x-3| > 5$$
$$x-3 < -5 \quad \text{or} \quad x-3 > 5$$
$$x < -2 \quad \text{or} \quad x > 8$$

Interval notation: $(-\infty, -2) \cup (8, \infty)$

Graph:

48.
$$|3(x+4)| < 6$$
$$|3x+12| < 6$$
$$-6 < \quad 3x+12 \quad < 6$$
$$-18 < \quad 3x \quad < -6$$
$$-6 < \quad x \quad < -2$$

Interval notation: $(-6, -2)$
Graph:

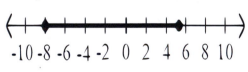

-10 -8 -6 -4 -2 0 2 4 6 8 10

54.
$$\left|\frac{3}{4}x+2\right| \geq 8$$

$$\frac{3}{4}x+2 \leq -8 \quad \text{or} \quad \frac{3}{4}x+2 \geq 8$$

$$4\left(\frac{3}{4}x+2\right) \leq 4(-8) \quad \text{or} \quad 4\left(\frac{3}{4}x+2\right) \geq 4(8)$$

$$3x+8 \leq -32 \quad \text{or} \quad 3x+8 \geq 32$$

$$3x \leq -40 \quad \text{or} \quad 3x \geq 24$$

$$x \leq -\frac{40}{3} \quad \text{or} \quad x \geq 8$$

Interval notation: $\left(-\infty, -\frac{40}{3}\right] \cup [8, \infty)$

Graph:

-10 -8 -6 -4 -2 0 2 4 6 8 10

60.
$$\left|\frac{3x+4}{5}\right| \qquad\qquad \leq \qquad\qquad 4$$
$$-4 \qquad \leq \frac{3x+4}{5} \leq \qquad 4$$
$$5(-4) \quad \leq 5\left(\frac{3x+4}{5}\right) \leq \quad 5(4)$$
$$-20 \qquad \leq 3x+4 \leq \qquad 20$$
$$-24 \qquad \leq 3x \leq \qquad 16$$
$$-8 \qquad \leq x \leq \qquad \frac{16}{3}$$

Interval notation: $\left[-8, \frac{16}{3}\right]$

Graph:

-10 -8 -6 -4 -2 0 2 4 6 8 10

1.3 Exercises

Find the imaginary part of each complex number.

2. The imaginary part of $5 + 3i$ is 3.

4. The imaginary part of $4i$ is 4.

6. The imaginary part of -4 is 0.

Calculate each of the following.

We will use the formulas $i^1 = i$, $i^2 = -1$, $i^3 = -i$, $i^4 = 1$. To calculate a higher power of i, first divide the exponent by 4. The remainder will be less than 4. Use one of these four formulas with the remainder as the exponent to calculate the answer.

8. $i^{223} = i^{220} \cdot i^3 = (i^4)^{55} \cdot i^3 = (1^{55}) \cdot i^3 = -i$.

12. $i^{-630} = i^{-632} \cdot i^2 = (i^4)^{-158} \cdot (-1) = 1^{-158} \cdot (-1) = -1$

14. $(7 - 5i) + (2 - 6i) = 7 - 5i + 2 - 6i = (7 + 2) + (-5i - 6i) = 9 - 11i$

18. $(13 + 12i) - (8 + 2i) = 13 + 12i - 8 - 2i = (13 - 8) + (12i - 2i) = 5 + 10i$

22. $-6(2 + 4i) = -6(2) - 6(4i) = -12 - 24i$

26. $(6 + i)(8 + 3i) = 48 + 18i + 8i + 3i^2 = 48 + 26i + 3(-1) = 45 + 26i$

32. $(6 - 2i)^2 = (6 - 2i)(6 - 2i) = 36 - 12i - 12i + 4i^2 = 36 - 24i + 4(-1) = 32 - 24i$

36. $(4 - 3i)^2(4 + 3i)^2 = ((4 - 3i)(4 + 3i))^2 = (16 + 12i - 12i - 9i^2)^2$

$$= (16 + 9)^2 = 25^2 = 625$$

40. $\dfrac{6 + i}{8 + 3i} = \dfrac{6 + i}{8 + 3i} \cdot \dfrac{8 - 3i}{8 - 3i} = \dfrac{(6 + i)(8 - 3i)}{(8 + 3i)(8 - 3i)} = \dfrac{48 - 18i + 8i - 3i^2}{64 - 24i + 24i - 9i^2}$

$$= \dfrac{48 - 10i - 3(-1)}{64 - 9(-1)} = \dfrac{51}{73} - \dfrac{10i}{73}$$

48. $\dfrac{(5 + 2i)(4 - i)}{3 + 2i} = \dfrac{20 - 5i + 8i - 2i^2}{3 + 2i} = \dfrac{22 + 3i}{3 + 2i} \cdot \dfrac{3 - 2i}{3 - 2i} = \dfrac{(22 + 3i)(3 - 2i)}{(3 + 2i)(3 - 2i)}$

$$= \dfrac{66 - 44i + 9i - 6i^2}{9 - 6i + 6i - 4i^2} = \dfrac{72}{13} - \dfrac{35i}{13}$$

1.4   Exercises

In problems 1 - 10, solve for x by factoring.

2. $x^2 - 3x - 10 = 0$

$(x - 5)(x + 2) = 0$

$x - 5 = 0$ or $x + 2 = 0$

$x = 5$ or $x = -2$

Solution set

$\{-2, 5\}$

8. $x(x - 4) = -4$

$x^2 - 4x = -4$

$x^2 - 4x + 4 = 0$

$(x - 2)(x - 2) = 0$

$x = 2$

Solution set $\{2\}$

In problems 11 - 20, solve for x by using completing the square.

Steps for completing the square:

1. Collect the variable terms on the left hand side, and the constant on the right hand side.

2. If the coefficient of x^2 is 1, then take half the coefficient of x and then square it. Add this number to both sides of the equation. If the coefficient of x^2 is not 1, then first divide both sides by the coefficient first.

3. Factor the left hand side. It should be a perfect square.

4. Take the square root of both sides. Remember the plus/minus on the right hand side.

5. Solve for x.

12. $x^2 - 6x + 3 = 0$

$x^2 - 6x = -3$

$x^2 - 6x + 9 = -3 + 9$

$(x - 3)^2 = 6$

$x - 2 = \pm\sqrt{6}$

$x = 2 \pm \sqrt{6}$

Solution set

$$\left\{ 2 - \sqrt{6}, 2 + \sqrt{6} \right\}$$

18. $2x = x^2 + 10x + 25$

$x^2 + 10x + 25 = 2x$

$x^2 + 8x + 25 = 0$

$x^2 + 8x = -25$

$x^2 + 8x + 16 = -25 + 16$

$(x + 4)^2 = -9$

$x + 4 = \pm\sqrt{-9}$

$x + 4 = \pm 3i$

$x = -3 \pm 3i$

Solution set

$$\{-3 - 3i, 3 + 3i\}$$

In problems 21 - 30, solve for x by using the quadratic formula.

In order to use the quadratic formula, first get a zero on the right hand side of the equation. The coefficient of x^2 is a. The coefficient of x is b. The constant term is c. The quadratic formula is

$$x = \frac{-b \pm \sqrt{b^2 - 4ac}}{2a}$$

22. $x^2 + 5x + 12 = 0$

   $a = 1, \ b = 5, \ c = 12$

   $x = \dfrac{-5 \pm \sqrt{5^2 - 4(1)(12)}}{2(1)}$

   $= \dfrac{-5 \pm \sqrt{25 - 48}}{2}$

   $= \dfrac{-5 \pm \sqrt{-23}}{2}$

   $= \dfrac{-5 \pm i\sqrt{23}}{2}$

   Solution set

   $\left\{ \dfrac{-5 \pm i\sqrt{23}}{2}, \dfrac{-5 \pm i\sqrt{23}}{2} \right\}$

30. $4x^2 = 3x + 2$

   $4x^2 - 3x - 2 = 0$

   $a = 4, \ b = -3, \ c = -2$

   $x = \dfrac{-(-3) \pm \sqrt{(-3)^2 - 4(4)(-2)}}{2(4)}$

   $= \dfrac{3 \pm \sqrt{9 + 32}}{8}$

   $= \dfrac{3 \pm \sqrt{41}}{8}$

   Solution set

   $\left\{ \dfrac{3 - \sqrt{41}}{8}, \dfrac{3 + \sqrt{41}}{8} \right\}$

1.5   Exercises

Find the solution set for each equation.

Here we first get a zero on the right hand side of the equation. If we cannot factor the left hand side, then we will use synthetic division. The Rational Roots Theorem says any rational root must be a factor of the constant term divided by a factor of the leading coefficient.

2. $x^3 + 2x^2 - x - 2 = 0$

   We   try $x = 1$:

   $\begin{array}{r|rrrr} 1 & 1 & 2 & -1 & -2 \\ & & 1 & 3 & 2 \\ \hline & 1 & 3 & 2 & 0 \end{array}$

   A   zero   in   the   box   means $x = 1$ is   a   solution:

   $(x - 1)(x^2 + 3x + 2) = 0$

   $(x - 1)(x + 2)(x + 1) = 0$

   Solution set

   $\{-2, -1, 1\}$

6. $x^3 - x^2 + 4x - 4 = 0$

   $(x^3 - x^2) + (4x - 4) = 0$

   $x^2(x - 1) + 4(x - 1) = 0$

   $(x^2 + 4)(x - 1) = 0$

   $x^2 + 4 = 0$ or $\qquad x - 1 = \ 0$

   $x = \pm 2i$ or $\qquad\qquad x = \ 1$

   Solution set

   $\{-2i, 2i, 1\}$

10. $x^3 - 8 = 0$

$(x - 2)(x^2 + 2x + 4) = 0$

$x - 2 = 0$ or $x^2 + 2x + 4 = 0$

$x = 2$ or $x = \dfrac{-2 \pm \sqrt{2^2 - 4(1)(4)}}{2(1)}$

$= \dfrac{-2 \pm \sqrt{-12}}{2}$

$= \dfrac{-2 \pm 2i\sqrt{3}}{2}$

$= -1 \pm i\sqrt{3}$

Solution set

$$\left\{2, -1 - i\sqrt{3}, -1 + i\sqrt{3}\right\}$$

12. $29x^2 = x^4 + 100$

$x^4 - 29x^2 + 100 = 0$

$(x^2 - 25)(x^2 - 4) = 0$

$(x + 5)(x - 5)(x + 2)(x - 2) = 0$

$x = -5, \ 5, \ -2, \ 2$

Solution set

$$\{-5, -2, 2, 5\}$$

18. $x^4 + 2x^3 = 5x^2 + 6x$

$x^4 + 2x^3 - 5x^2 - 6x = 0$

$x(x^3 + 2x^2 - 5x - 6) = 0$

Now synthetic division with $x = 2$:

```
2 | 1   2   -5   -6
  |     2    8    6
  ------------------
    1   4   -3 | 0
```

This works, so we have

$x(x - 2)(x^2 + 4x - 3) = 0$

Now the quadratic formula:

$a = 1, \ b = 4, \ c = -3$

$x = \dfrac{-4 \pm \sqrt{4^2 - 4(1)(-3)}}{2(1)}$

$= \dfrac{-4 \pm \sqrt{28}}{2}$

$= \dfrac{-4 \pm 2\sqrt{7}}{2}$

$= -2 \pm \sqrt{7}$

The roots are:

$x = 0, 2, -2 \pm \sqrt{7}$

Solution set

$$\left\{0, 2, -2 - \sqrt{7}, -2 + \sqrt{7}\right\}$$

30. $4x^3 - x = 4x^4 - 9x^2 + 2$

$4x^4 - 4x^3 - 9x^2 + x + 2 = 0$

We will try $x = -1$:

```
-1 | 4   -4   -9    1    2
   |      -4    8    1   -2
   -------------------------
     4   -8   -1    2 | 0
```

This works, so we get

$(x + 1)(4x^3 - 8x^2 - x + 2) = 0$

We now use factoring by grouping:

$(x + 1)((4x^3 - 8x^2) - (x - 2)) = 0$

$(x + 1)(4x^2(x - 2) - (x - 2)) = 0$

$(x + 1)(x - 2)(4x^2 - 1) = 0$

$(x + 1)(x - 2)(2x + 1)(2x - 2) = 0$

$x = -1, \ 2, \ -\dfrac{1}{2}, \ \dfrac{1}{2}$

Solution set

$$\left\{-1, -\dfrac{1}{2}, \dfrac{1}{2}, 2\right\}$$

1.6   Exercises

Find the solution set for each equation.

For these problems, first find the least common denominator. Next multiply both sides of the equation by the LCD. Then solve for x. Remember to always check your answer in the original equation.

4. $\dfrac{2x}{x-4} = 2 + \dfrac{8}{x-4}$

$(x-4)\left(\dfrac{2x}{x-4}\right) = (x-4)\left(2 + \dfrac{8}{x-4}\right)$

$2x = 2(x-4) + 8$

$2x = 2x - 8 + 8$

$2x = 2x$

If we subtract $2x$ from both sides of the equation, we get $0 = 0$, so any real number is a solution except for $x = 4$.

Solution set

$\{4\}$

8. $\dfrac{x-2}{x+3} = \dfrac{x+2}{x-1}$

$(x+3)(x-1)\left(\dfrac{x-2}{x+3}\right) = (x+3)(x-1)\left(\dfrac{x+2}{x-1}\right)$

$(x-1)(x-2) = (x+3)(x+2)$

$x^2 - 3x + 2 = x^2 + 5x + 6$

$-3x + 2 = 5x + 6$

$-3x = 5x + 4$

$-8x = 4$

$x = -\dfrac{1}{2}$

This value of x does not cause either denominator to equal 0, so it solves the equation.

Solution set

$\left\{-\dfrac{1}{2}\right\}$

12. $\dfrac{x}{2x+1} + \dfrac{3}{2x-1} = \dfrac{5x+11}{4x^2-1}$

$(2x+1)(2x-1)\left(\dfrac{x}{2x+1} + \dfrac{3}{2x-1}\right) = (2x+1)(2x-1)\left(\dfrac{5x+11}{4x^2-1}\right)$

$x(2x-1) + 3(2x+1) = 5x+11$

$2x^2 - x + 6x + 3 = 5x + 11$

$2x^2 - 8 = 0$

$2(x^2 - 4) = 0$

$2(x+2)(x-2) = 0$

$x = -2,\ 2$

Neither value of x causes either denominator to equal 0, so they both solve the equation.

Solution set

$\{-2, 2\}$

18. $\dfrac{2x-1}{x-3} + \dfrac{3x-1}{x-2} = \dfrac{-5}{x^2-5x+6}$

$(x-3)(x-2)\left(\dfrac{2x-1}{x-3} + \dfrac{3x-1}{x-2}\right) = (x-3)(x-2)\left(\dfrac{-5}{x^2-5x+6}\right)$

$(x-2)(2x-1) + (x-3)(3x-1) = -5$

$2x^2 - x - 4x + 2 + 3x^2 - x - 9x + 3 = 5$

$5x^2 - 15x + 5 = 5$

$5x^2 - 15x = 0$

$5x(x-3) = 0$

$x = 0,\ 3$

When $x = 3$, the denominators of the first and last fraction are both zero, so this is an extraneous solution. Therefore the only solution to the equation is $x = 0$.

Solution set

$\{-2, 2\}$

1.7   Exercises

Solve each polynomial inequality and graph the solution on the number line.

To solve these inequalities, we get zero on the right hand side of the equation. Then factor the left hand side. Set each factor equal to zero. The solutions to these equations will be the endpoints of the intervals of solution to the inequality. Then choose test points between each solution to see if the inequality is satisfied.

2. $x^2 - 3x - 10 < 0$

$(x - 5)(x + 2) < 0$

$x = -2,\ 5$

Test points: $x = -3,\ 0,\ 6$

$x = -3$ does not work

$x = 0$ does work

$x = 6$ does not work

The solution set is $-2 < x < 5$

Interval notation: $(-2, 5)$

Graph:

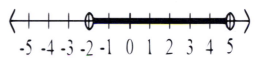

8. $x(x - 4) \geq -4$

$x^2 - 4x \geq -4$

$x^2 - 4x + 4 \geq 0$

$(x - 2)(x - 2) \geq 0$

$x = 2,\ 2$

Test points: $x = 0,\ 3$

$x = 0$ works

$x = 3$ works

$x = 2$ works

The solution set is all real numbers.

Interval notation: $(-\infty, \infty)$

Graph:

12. $x^3 + 2x^2 > x + 2$

$x^3 + 2x^2 - x - 2 > 0$

$(x^3 + 2x^2) - (x + 2) > 0$

$x^2(x + 2) - (x + 2) > 0$

$(x + 2)(x^2 - 1) > 0$

$(x + 2)(x + 1)(x - 1) > 0$

$x = -2, -1, 1$

Test points: $x = -3, -1.5, 0, 2$

$x = -3$ does not work

$x = -1.5$ works

$x = 0$ does not work

$x = 2$ works

Interval notation: $(-2, -1) \cup (1, \infty)$
Graph:

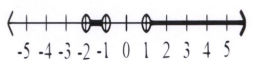

18. $x^3 - 3x < 2$

$x^3 - 3x - 2 < 0$

$(x + 1)^2(x - 2) < 0$

$x = -1, -1, 2$

Test points: $x = -2, 0, 3$

$x = -2$ works

$x = 0$ works

$x = 3$ does not work

$x = -1$ does not work

Interval notation: $(-\infty, -1) \cup (-1, 2)$
Graph:

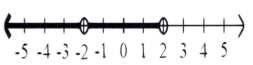

24. $\dfrac{(x - 2)(x + 1)}{x - 3} < 0$

$x = -1, 2, 3$

Test points: $x = -2, 0, 2.5, 4$

$x = -2$ works

$x = 0$ does not work

$x = 2.5$ works

$x = 4$ does not work

Interval notation: $(-\infty, -1) \cup (2, 3)$
Graph:

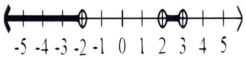
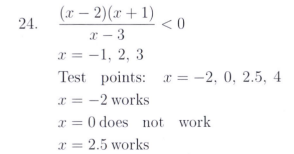

28. $\dfrac{5}{x-3} < \dfrac{3}{x+1}$

$\dfrac{5}{x-3} - \dfrac{3}{x+1} < 0$

$\dfrac{5(x+1)}{(x+1)(x-3)} - \dfrac{3(x-3)}{(x+1)(x-3)} < 0$

$\dfrac{2x+14}{(x+1)(x-3)} < 0$

$\dfrac{2(x+7)}{(x+1)(x-3)} < 0$

$x = -7, -1, 3$

Test points: $x = -8, -2, 0, 4$

$x = -8$ works

$x = -2$ does not work

$x = 0$ works

$x = 4$ does not work

Interval notation: $(-\infty, -7) \cup (-1, 3)$

Graph:

-10 -8 -6 -4 -2 0 2 4 6 8 10

1.8   Exercises

Solve each equation for x.

When solving radical equations, first isolate the radical (or a radical) on one side of the equation. Then eliminate it by raising both sides of the equation to the same power. Be sure to check the final answer(s) to make sure they work in the original equation.

4. $\sqrt{5x+11} = 6$

$\left(\sqrt{5x+11}\right)^2 = 6^2$

$5x + 11 = 36$

$5x = 25$

$x = 5$

This solution works in the original equation.

Solution set

$\{5\}$

10. $\sqrt[3]{2x+1} = 5$

$\left(\sqrt[3]{2x+1}\right)^3 = 5^3$

$2x + 1 = 125$

$2x = 124$

$x = 62$

This solution works in the original equation.

Solution set

$\{62\}$

14. $\sqrt{2x-5} = x-2$

$\left(\sqrt{2x-5}\right)^2 = (x-2)^2$

$2x-5 = x^2 - 4x + 4$

$x^2 - 6x + 9 = 0$

$(x-3)(x-3) = 0$

$x = 3$

This solution works.

Solution set

$\{3\}$

18. $2x-1 = \sqrt{2x+5}$

$(2x-1)^2 = \left(\sqrt{2x+5}\right)^2$

$4x^2 - 4x + 1 = 2x + 5$

$4x^2 - 6x - 4 = 0$

$2(2x+1)(x-2) = 0$

$x = -\dfrac{1}{2}, \ 2$

Only $x = 2$ works.

Solution set

$\{2\}$

When solving an equation with two radicals, first isolate one radical on one side of the equation, then square both sides,

22. $\sqrt{2x+5} + \sqrt{x+2} = 5$

$\sqrt{2x+5} = 5 - \sqrt{x+2}$

$\left(\sqrt{2x+5}\right)^2 = \left(5 - \sqrt{x+2}\right)^2$

$2x+5 = (5 - \sqrt{x+2})(5 - \sqrt{x+2})$

$2x+5 = 25 - 10\sqrt{x+2} + (x+2)$

$x - 22 = -10\sqrt{x+2}$

$(x-22)^2 = \left(-10\sqrt{x+2}\right)^2$

$x^2 - 44x + 484 = 100(x+2)$

$x^2 - 44x + 484 = 100x + 200$

$x^2 - 144x + 284 = 0$

$(x-142)(x-2) = 0$

$x = 2, \ 142$

$x = 2$ works

$x = 142$ does not work.

Solution set

$\{2\}$

28. $\sqrt{x} = \sqrt{x+24} - 2$

$\left(\sqrt{x}\right)^2 = \left(\sqrt{x+24} - 2\right)^2$

$x = (\sqrt{x+24} - 2)(\sqrt{x+24} - 2)$

$x = (x+24) - 4\sqrt{x+24} + 4$

$x = x + 28 - 4\sqrt{x+24}$

$4\sqrt{x+24} = 28$

$\sqrt{x+24} = 7$

$\left(\sqrt{x+24}\right)^2 = 7^2$

$x + 24 = 49$

$x = 25$

This solution works.

Solution set

$\{25\}$

Chapter 2

Functions

2.1 Exercises

Determine whether the correspondence is a function.

Remember that a function is a relation that satisfies the condition that every domain value gets mapped to exactly one range value.

$$
\begin{array}{rcl}
a & \to & 5 \\
2. \quad b & \to & 2 \\
c & \to & -1
\end{array}
$$

This is a function because every entry on the left is unique, and maps to exactly one entry on the right.

$$
\begin{array}{rcl}
q & \to & x \\
4. \quad b & \to & 2 \\
q & \to & -1
\end{array}
$$

This is not a function because the letter q shows up twice on the left, and maps to a different value on the right each time.

8. Cars in a parking lot $\to$ The color of the car $\to$ A set of colors

This does not describe a function because a car might have more than one color.

Determine whether each of the following are functions. Then identify the domain and range.

12. $\{(4,2),(5,3),(6,4),(7,5),(8,6)\}$

This is a function. Each x value is unique. The domain is $\{4,5,6,7,8\}$ and the range is $\{2,3,4,5,6\}$.

16. $\{(4,1),(5,2),(6,3),(5,4),(4,5)\}$

This is not a function because the x values 4 and 5 both appear twice, and each time are mapped to a different y value. The domain of this relation is $\{4,5,6\}$ and the range is $\{1,2,3,4,5\}$.

18. Given that $f(x) = 4x + 5$, calculate each of the following:
   a) $f(-1)$ b) $f(-2)$
   c) $f(0)$ d) $f(1)$

We have the following:

a) $f(-1) = 4(-1) + 5 = 1$ b) $f(-2) = 4(-2) + 5 = -3$
c) $f(0) = 4(0) + 5 = 5$ d) $f(1) = 4(1) + 5 = 9$

20. Given that $g(x) = 5x^2 + 4x$, calculate each of the following:
   a) $g(2)$ b) $g(-2)$
   c) $g(1)$ d) $g(-1)$

We have the following:

a) $g(2) = 5(2)^2 + 4(2) = 28$ b) $g(-2) = 5(-2)^2 + 4(-2) = 12$
c) $g(1) = 5(1)^2 + 4(1) = 9$ d) $g(-1) = 5(-1)^2 + 4(-1) = 1$

22. Given that $h(x) = \dfrac{x^2 - 1}{x + 2}$, calculate each of the following:

a) $h(3)$ b) $h(2)$
c) $h(-1)$ d) $h(-2)$

We have the following:

a) $h(3) = \dfrac{3^2 - 1}{3 + 2} = \dfrac{8}{5}$ b) $h(2) = \dfrac{2^2 - 1}{2 + 2} = \dfrac{3}{4}$

c) $h(-1) = \dfrac{(-1)^2 - 1}{(-1) + 2} = 0$ d) $h(-2) = \dfrac{(-2)^2 - 1}{(-2) + 2} = \dfrac{3}{0}$, which does not exist

In problem 24, the graph of $f(x)$ is given. Estimate $f(-1)$, $f(0)$, and $f(1)$.

24.

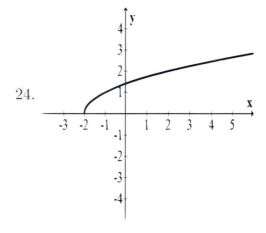

We can visually estimate: $f(-1) \approx 1$, $f(0) \approx 1.5$, $f(1) \approx 1.7$

In exercises 27 - 34, find the domain of the function. Do not use a grapher.

28. $f(x) = 4x + 5$

This function has no domain restrictions. The domain of f is all real numbers.

32. $f(x) = \dfrac{x-1}{x+2}$

We cannot use the value $x = -2$ in this function. The domain is all real numbers except -2. This can be written in interval notation as $(-\infty, -2) \cup (-2, \infty)$ or in set builder notation as $\{x \in \Re \mid x \neq -2\}$.

34. $g(x) = \sqrt{3x - 6}$

For this function, the radicand cannot be negative. Therefore we must have

$$
\begin{aligned}
3x - 6 &\geq 0 \\
3x &\geq 6 \\
x &\geq 2
\end{aligned}
$$

Therefore the domain restriction is $x \geq 2$. This can be written in interval notation as $[2, \infty)$.

In exercises 35 - 42, use a grapher to graph the function. Then visually estimate the domain and range.

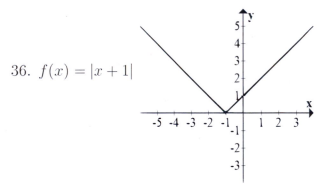

36. $f(x) = |x + 1|$

The domain of this function is all real numbers, and the range is restricted to $y \geq 0$.

40. $g(x) = 3 - x^2$

The domain of this function is all real numbers, and the range is restriced to $y \leq 3$.

42. $g(x) = (x-1)^2 + 3$

The domain of this function is all real numbers, and the range is restricted to $y \geq 3$.

In exercises 43-48, determine whether the graph is a graph of a function.

44.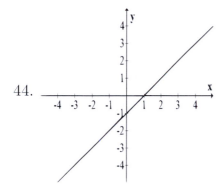

This is the graph of a function because it passes the vertical line test.

46.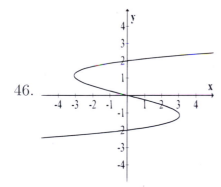

This is not the graph of a function because it fails the vertical line test.

2.2 Exercises

Complete a table of values for each of the following functions, using the x values as suggested. Then plot the points on a Cartesian plane and connect the points. Finally, identify any intercepts the graph may have.

2. $f(x) = 3x - 4$, $x = \{-2, -1, 0, 1, 2\}$

x	y
-2	-10
-1	-7
0	-4
1	-1
2	2

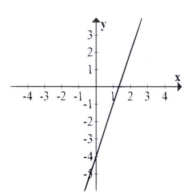

6. $f(x) = |x + 1|$, $x = \{-3, -2, -1, 0, 1, 2\}$

x	y
-3	2
-2	1
-1	0
0	1
1	2
2	3

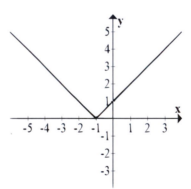

12. $f(x) = \sqrt{4 - x}$, $x = \{-5, 0, 3, 4\}$

x	y
-5	3
0	2
3	1
4	0

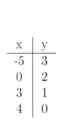
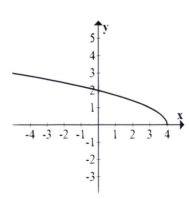

16. $g(x) = (x-1)^2 + 3$, $x = \{-1, 0, 1, 2, 3\}$

x	y
-1	7
0	4
1	3
2	4
3	7

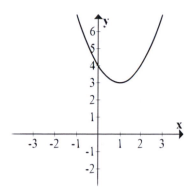

18. $g(x) = x^3 - 2x^2 - 8x$, $x = \{-4, -2, 0, 2, 4, 6\}$

x	y
-4	-64
-2	0
0	0
2	-16
4	0
6	120

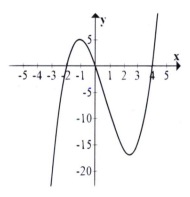

2.3   Exercises

In exercises 1-6, identify any maxima and minima. Then determine the values of x for which each function is increasing and decreasing.

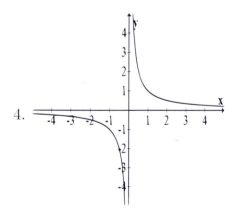

4.

This graph does not have any maximum or minimum points. The graph is decreasing for all x, except at $x = 0$, where the function does not exist.

6.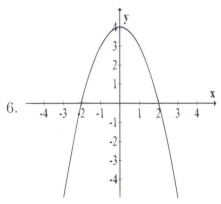

This graph has a maximum value at $x = 0$. The maximum value of the function is $y = 4$. The function increases when $x < 0$ and decreases when $x > 0$.

For each of the functions in problems 7 - 14, create a graph. Then identify any relative or absolute maxima and minima. Finally, identify where the function is increasing and decreasing.

8. $n(x) = 4 - 2x$

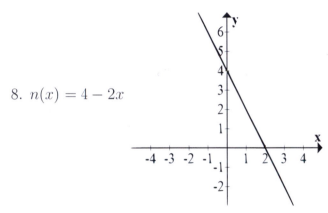

This graph does not have any maximum or minimum points. The graph is decreasing for all values of x.

12. $h(x) = x^2 - 3x + 2$

This graph has a minimum value at $x = 1.5$. The minimum value is $y = -0.25$. The graph is decreasing when $x < 1.5$ and increasing when $x > 1.5$.

For each function in problems 15 - 20, calculate the average rate of change:

a. from $x = 1$ to $x = 2$ *b. from $x = 1$ to $x = 1 + h$*

For both of these problems we will use the formula

$$\text{Average Rate of Change } = \frac{f(b) - f(a)}{b - a}$$

For part a we have $a = 1$, $b = 2$, and for part b we have $a = 1$, $b = 1 + h$.

16. $h(x) = 3x + 2$

a. $h(a) = h(1) = 3(1) + 2 = 5$

$h(b) = h(2) = 3(2) + 2 = 8$

$\dfrac{h(b) - h(a)}{b - a} = \dfrac{8 - 5}{2 - 1} = 3$

b. $h(a) = h(1) = 3(1) + 2 = 5$

$h(b) = h(1 + h) = 3(1 + h) + 2 = 3 + 3h + 2 = 3h + 5$

$\dfrac{h(b) - h(a)}{b - a} = \dfrac{(3h + 5) - 5}{(1 + h) - 1} = \dfrac{3h}{h} = 3$

18. $k(x) = x^2 - 3x + 2$

a. $k(a) = k(1) = 1^2 - 3(1) + 2 = 0$

$k(b) = k(2) = 2^2 - 3(2) + 2 = 0$

$\dfrac{k(b) - k(a)}{b - a} = \dfrac{0 - 0}{2 - 1} = 0$

b. $k(a) = k(1) = 1^2 - 3(1) + 2 = 0$

$k(b) = k(1 + h) = (1 + h)^2 - 3(1 + h) + 2 = (1 + 2h + h^2) - (3 + 3h) + 2 = h^2 - h$

$\dfrac{k(b) - k(a)}{b - a} = \dfrac{(h^2 - h) - 0}{(1 + h) - 1} = \dfrac{h(h - 1)}{h} = h - 1$

2.4   Exercises

In exercises 1-10, calculate $(f + g)(x)$ and $(f - g)(x)$.

Here we use the formulas $(f + g)(x) = f(x) + g(x)$ and $(f - g)(x) = f(x) - g(x)$.

2. $f(x) = 4x - 3$
    $g(x) = 3x + 2$

$(f + g)(x) = f(x) + g(x) = (4x - 3) + (3x + 2) = 4x - 3 + 3x + 2 = 7x - 1$

$(f - g)(x) = f(x) - g(x) = (4x - 3) - (3x + 2) = 4x - 3 - 3x - 2 = x - 5$

6. $f(x) = x^2 + 4$

$g(x) = 3x - 4$

$(f + g)(x) = f(x) + g(x) = (x^2 + 4) + (3x - 4) = x^2 + 4 + 3x - 4 = x^2 + 3x$

$(f - g)(x) = f(x) - g(x) = (x^2 + 4) - (3x - 4) = x^2 + 4 - 3x + 4 = x^2 - 3x + 8$

10. $f(x) = x^3 + 2x^2$

$g(x) = x^2 + 4$

$(f + g)(x) = f(x) + g(x) = (x^3 + 2x^2) + (x^2 + 4) = x^3 + 2x^2 + x^2 + 4 = x^3 + 3x^2 + 4$

$(f - g)(x) = f(x) - g(x) = (x^3 + 2x^2) - (x^2 + 4) = x^3 + 2x^2 - x^2 - 4 = x^3 + x^2 - 4$

In exercises 11-20, calculate $(f \cdot g)(x)$ and $\left(\dfrac{f}{g}\right)(x)$.

In these problems we use the formulas $(f \cdot g)(x) = f(x) \cdot g(x)$ and $\left(\dfrac{f}{g}\right)(x) = \dfrac{f(x)}{g(x)}$.

12. $f(x) = 5x - 4$

$g(x) = 4x + 3$

$(f \cdot g)(x) = f(x) \cdot g(x) = (5x - 4)(4x + 3) = 20x^2 + 15x - 16x - 12 = 20x^2 - x - 12$

$\left(\dfrac{f}{g}\right)(x) = \dfrac{f(x)}{g(x)} = \dfrac{5x - 4}{4x + 3}$

16. $f(x) = x^2 + 3$

$g(x) = 6x - 2$

$(f \cdot g)(x) = f(x) \cdot g(x) = (x^2 + 3)(6x - 2) = 6x^3 - 2x^2 + 18x - 6$

$\left(\dfrac{f}{g}\right)(x) = \dfrac{f(x)}{g(x)} = \dfrac{x^2 + 3}{6x - 2}$

20. $f(x) = 2x^3 + 5$

$g(x) = x^2 + 4$

$(f \cdot g)(x) = f(x) \cdot g(x) = (2x^3 + 5)(x^2 + 4) = 2x^5 + 8x^3 + 5x^2 + 20$

$\left(\dfrac{f}{g}\right)(x) = \dfrac{f(x)}{g(x)} = \dfrac{2x^3 + 5}{x^2 + 4}$

In exercises 21-30, calculate $(f+g)(2)$ and $(f-g)(2)$.

22. $f(x) = 4x - 3$
    $g(x) = 3x + 2$

   $(f+g)(2) = f(2) + g(2) = (4(2)-3) + (3(2)+2) = 5+8 = 13$
   $(f-g)(2) = f(2) - g(2) = (4(2)-3) - (3(2)+2) = 5-8 = -3$

26. $f(x) = x^2 + 4$

   $g(x) = 3x - 4$

   $(f+g)(2) = f(2) + g(2) = (2^2+4) + (3(2)-4) = 8+2 = 10$
   $(f-g)(2) = f(2) - g(2) = (2^2+4) - (3(2)-4) = x^2+4-3x+4 = x^2-3x+8$

30. $f(x) = x^3 + 2x^2$

   $g(x) = x^2 + 4$

   $(f+g)(2) = f(2) + g(2) = (2^3 + 2(2)^2) + ((2)^2+4) = 16+8 = 24$
   $(f-g)(2) = f(2) - g(2) = (2^3 + 2(2)^2) - ((2)^2+4) = 16-8 = 8$

In exercises 31-40, calculate $(f \cdot g)(3)$ and $(f/g)(3)$.

32. $f(x) = 5x - 4$

   $g(x) = 4x + 3$

   $(f \cdot g)(3) = f(3) \cdot g(3) = (5(3)-4)(4(3)+3) = (11)(15) = 165$

   $\left(\dfrac{f}{g}\right)(3) = \dfrac{5(3)-4}{4(3)+3} = \dfrac{11}{15}$

36. $f(x) = x^2 + 3$

   $g(x) = 6x - 2$

   $(f \cdot g)(3) = f(3) \cdot g(3) = (3^2+3)(6(3)-2) = (12)(16) = 192$

   $\left(\dfrac{f}{g}\right)(3) = \dfrac{f(3)}{g(3)} = \dfrac{3^2+3}{6(3)-2}\dfrac{3}{4}$

40. $f(x) = 2x^3 + 5$

   $g(x) = x^2 + 4$

   $(f \cdot g)(3) = f(3) \cdot g(3) = (2 \cdot 3^3 + 5)(3^2+4) = (23)(13) = 299$

   $\left(\dfrac{f}{g}\right)(3) = \dfrac{f(3)}{g(3)} = \dfrac{2(3)^2+5}{3^2+4} = \dfrac{23}{13}$

In exercises 41-50, calculate $(f \circ g)(x)$ and $(g \circ f)(x)$.

Remember the definition of composition: $(f \circ g)(x) = f(g(x))$.

42. $f(x) = 4x - 3$

$g(x) = 3x + 2$

$(f \circ g)(x) = f(g(x)) = f(3x + 2) = 4(3x + 2) - 3 = 12x + 8 - 3 = 12x + 5$

$(g \circ f)(x) = g(f(x)) = g(4x - 3) = 3(4x - 3) + 2 = 12x - 9 + 2 = 12x - 7$

46. $f(x) = 6x - 2$

$g(x) = 3x + 1$

$(f \circ g)(x) = f(g(x)) = f(3x + 1) = 6(3x + 1) - 2 = 18x + 6 - 2 = 18x + 4$

$(g \circ f)(x) = g(f(x)) = g(6x - 2) = 3(6x - 2) + 1 = 18x - 6 + 1 = 18x - 5$

50. $f(x) = x^2 + 4$

$g(x) = 3x - 4$

$(f \circ g)(x) = f(g(x)) = f(3x - 4) = (3x - 4)^2 + 4 = (3x - 4)(3x - 4) + 4$

$\quad = 9x^2 - 12x - 12x + 16 + 4 = 9x^2 - 24x + 20$

$(g \circ f)(x) = g(f(x)) = g(x^2 + 4) = 3(x^2 + 4) - 4 = 3x^2 + 12 - 4 = 3x^2 + 8$

In exercises 51-60, calculate $(f \circ g)(2)$ and $(g \circ f)(2)$.

52. $f(x) = 4x - 3$

$g(x) = 3x + 2$

$(f \circ g)(2) = f(g(2)) = f(3(2) + 2) = f(8) = 4(8) - 3 = 29$

$(g \circ f)(2) = g(f(2)) = g(4(2) - 3)g(5) = 3(5) + 2 = 17$

56. $f(x) = 6x - 2$

$g(x) = 3x + 1$

$(f \circ g)(2) = f(g(2)) = f(3(2) + 1) = f(7) = 6(7) - 2 = 40$

$(g \circ f)(2) = g(f(2)) = g(6(2) - 2) = g(10) = 3(10) + 1 = 31$

60. $f(x) = x^2 + 4$

$g(x) = 3x - 4$

$(f \circ g)(2) = f(g(2)) = f(3(2) - 4) = f(2) = 2^2 + 4 = 8$

$(g \circ f)(2) = g(f(2)) = g(2^2 + 4) = g(8) = 3(8) - 4 = 20$

2.5   Exercises

For each of the graphs in 1-6 determine whether the function is one to one:

2.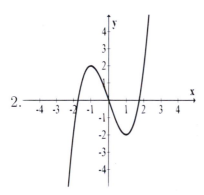

This function is not one-to-one. If we draw a horizontal line through the middle portion of the graph (say, at $y = 1$), then it crosses the graph three times. Therefore it fails the horizontal line test.

6.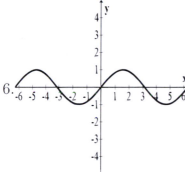

This function is not one-to-one. If we draw a horizontal line through the middle portion of the graph (say, at $y = 0.5$), then it crosses the graph four times. Therefore it fails the horizontal line test.

For problems 7-16 verify that $f(x)$ and $g(x)$ are inverse functions by showing that $f(g(x)) = g(f(x)) = x$:

10. $f(x) = 2x - 1$

$$g(x) = \frac{x + 1}{2}$$

$$f(g(x)) = f\left(\frac{x + 1}{2}\right) = 2\left(\frac{x + 1}{2}\right) - 1 = (x + 1) - 1 = x$$

$$g(f(x)) = g(2x - 1) = \frac{(2x - 1) + 1}{2} = \frac{2x}{2} = x$$

14. $f(x) = 2x^3 + 5$

$$g(x) = \sqrt[3]{\frac{x-5}{2}}$$

$$f(g(x)) = f\left(\sqrt[3]{\frac{x-5}{2}}\right) = 2\left(\sqrt[3]{\frac{x-5}{2}}\right)^3 + 5 = 2\left(\frac{x-5}{2}\right) + 5 = (x-5) + 5 = x$$

$$g(f(x)) = g(2x^3 + 5) = \sqrt[3]{\frac{(2x^3+5)-5}{2}} = \sqrt[3]{\frac{2x^3}{3}} = \sqrt[3]{x^3} = x$$

In problems 17-30, find the inverse of each function:

In order to find the inverse of the function $f(x)$, follow these steps:

1. Replace $f(x)$ with y.

2. Switch the x and y variables.

3. Solve the new equation for y.

4. Replace y with $f^{-1}(x)$.

18. $f(x) = 5x - 3$

$y = 5x - 3$

$x = 5y - 3$

$5y = x + 3$

$y = \dfrac{x+3}{5}$

$f^{-1}(x) = \dfrac{x+3}{5}$

22. $f(x) = 2x^3 - 7$

$y = 2x^3 - 7$

$x = 2y^3 - 7$

$2y^3 = x + 7$

$y^3 = \dfrac{x+7}{2}$

$y = \sqrt[3]{\dfrac{x+7}{2}}$

$f^{-1}(x) = \sqrt[3]{\dfrac{x+7}{2}}$

26. $f(x) = \sqrt[3]{4x+1}$

$\qquad y = \sqrt[3]{4x+1}$

$\qquad x = \sqrt[3]{4y+1}$

$\qquad x^3 = 4y+1$

$\qquad 4y = x^3 - 1$

$\qquad y = \dfrac{x^3 - 1}{4}$

$\qquad f^{-1}(x) = \dfrac{x^3 - 1}{4}$

30. $f(x) = \dfrac{2x-4}{x-3}$

$\qquad y = \dfrac{2x-4}{x-3}$

$\qquad x = \dfrac{2y-4}{y-3}$

$\qquad x(y-3) = 2y-4$

$\qquad xy - 3x = 2y - 4$

$\qquad xy = 2y + 3x - 4$

$\qquad xy - 2y = 3x - 4$

$\qquad y(x-2) = 3x - 4$

$\qquad y = \dfrac{3x-4}{x-2}$

$\qquad f^{-1}(x) = \dfrac{3x-4}{x-2}$

Chapter 3

Polynomial and Rational Functions

3.1 Exercises

Identify the slope and y-intercept of each of the following functions. Then calculate the x-intercept and draw a graph of the line.

Remember that the equation of a line in slope-intercept form is $y = mx + b$. Here m represents the slope and b represents the y-intercept.

2. $f(x) = 2x - 1$

   We have $m = 2$ and $b = -1$, so the slope is 2 and the y-intercept is -1.

10. $f(x) = -2x + 1$

    We have $m = -2$ and $b = 1$, so the slope is -2 and the y-intercept is 1.

In exercises 11 - 20, calculate the slope of a line passing through each pair of points.

Given two points (x_1, y_1) and (x_2, y_2), we calculate the slope using the formula

$$m = \frac{y_2 - y_1}{x_2 - x_1}$$

If $x_1 = x_2$ then the denominator becomes zero. In this case we say we have no slope.

12. $(1, 2)$ and $(2, 3)$

    We have $x_1 = 1$, $y_1 = 2$, $x_2 = 2$, and $y_2 = 3$. Therefore $m = \dfrac{3 - 2}{2 - 1} = 1$

16. $(1, 2)$ and $(4, -1)$

    We have $x_1 = 1$, $y_1 = 2$, $x_2 = 4$, and $y_2 = -1$. Therefore $m = \dfrac{-1 - 2}{4 - 1} = -\dfrac{3}{4}$

20. $(3, 2)$ and $(6, 2)$

We have $x_1 = 3$, $y_1 = 2$, $x_2 = 6$, and $y_2 = 2$. Therefore $m = \dfrac{2 - 2}{6 - 3} = 0$

In exercises 21-30, find the equation of the line passing through each pair of points.

Given two points (x_1, y_1) and (x_2, y_2), if we wish to find the equation of the line passing through these points, we use the following steps:

1. Calculate the slope of the line using the formula $m = \dfrac{y_2 - y_1}{x_2 - x_1}$.

2. Substitute x_1, y_1, and m into the point-slope form $y - y_1 = m(x - x_1)$.

3. Solve the previouse equation for y.

If $x_1 = x_2$ then we have a vertical line, and the equation of the line is $x = x_1$.

22. $(1, 2)$ and $(2, 3)$

We have $x_1 = 1$, $y_1 = 2$, $x_2 = 2$, and $y_2 = 3$.

First the slope: $m = \dfrac{y_2 - y_1}{x_2 - x_1} = \dfrac{3 - 2}{2 - 1} = 1$

Next the slope-intercept form: $y - 2 = 1(x - 1)$

Now we solve for y: $y - 2 = x - 1$ becomes $y = x + 1$.

26. $(1, 2)$ and $(4, -1)$

We have $x_1 = 1$, $y_1 = 2$, $x_2 = 4$, and $y_2 = -1$.

First the slope: $m = \dfrac{y_2 - y_1}{x_2 - x_1} = \dfrac{-1 - 2}{4 - 1} = -\dfrac{3}{4}$

Next the slope-intercept form: $y - 2 = -\dfrac{3}{4}(x - 1)$

Now we solve for y: $y - 2 = -\dfrac{3}{4}(x - 1)$ becomes $y = -\dfrac{3}{4}x + \dfrac{11}{4}$.

30. $(3, 2)$ and $(6, 2)$

We have $x_1 = 3$, $y_1 = 2$, $x_2 = 6$, and $y_2 = 2$.

First the slope: $m = \dfrac{y_2 - y_1}{x_2 - x_1} = \dfrac{2 - 2}{6 - 3} = 0$

Next the slope-intercept form: $y - 2 = 0(x - 3)$

Now we solve for y: $y - 2 = 0(x - 3)$ becomes $y = 2$.

In exercises 31-40, find the equation of the line parallel to the given line, passing through the given point.

In order to find the equation of a line passing through the point (x_1, y_1) and parallel to the line $y = mx + b$, use the same slope, then the point-slope formula $y - y_1 = m(x - x_1)$.

32. $y = x + 2$ through $(2, 3)$

We have $m = 1$, $x_1 = 2$, and $y_1 = 3$. Therefore the equation of the line becomes

$$y - 3 = 1(x - 2)$$

$$y - 3 = x - 2$$

$$y = x + 1$$

38. $y = -2x - 3$ through $(-4, 5)$

We have $m = -2$, $x_1 = -4$, and $y_1 = 5$. Therefore the equation of the line becomes

$$y - 5 = -2(x - (-4))$$

$$y - 5 = -2(x + 4)$$

$$y - 5 = -2x - 8$$

$$y = -2x - 3$$

40. $y = 3x + 2$ through $(6, 2)$

We have $m = 3$, $x_1 = 6$, and $y_1 = 2$. Therefore the equation of the line becomes

$$y - 2 = 3(x - 6)$$

$$y - 2 = 3x - 18$$

$$y = 3x - 16$$

In exercises 41-50, find the equation of the line perpendicular to the given line, passing through the given point.

In order to find the equation of a line passing through the point (x_1, y_1) and perpendicular to the line $y = mx + b$, use the new slope $m_1 = -\dfrac{1}{m}$, then the point-slope formula $y - y_1 = -\dfrac{1}{m}(x - x_1)$. If the old slope is zero, then the new slope is undefined. This will be a vertical line with equation $x = x_1$.

42. $y = x + 2$ through $(2, 3)$

We have $m = 1$, $x_1 = 2$, and $y_1 = 3$. The new slope is $m_1 = -\dfrac{1}{1} = -1$.

The equation of the line becomes

$y - 3 = 1(x - 2)$

$y - 3 = x - 2$

$y = x + 1$

48. $y = -2x - 3$ through $(-4, 5)$

We have $m = -2$, $x_1 = -4$, and $y_1 = 5$. The new slope is $m_1 = -\dfrac{1}{-2} = \dfrac{1}{2}$.

The equation of the line becomes

$y - 5 = \dfrac{1}{2}(x - (-4))$

$y - 5 = \dfrac{1}{2}(x + 4)$

$y - 5 = \dfrac{1}{2}x + 2$

$y = \dfrac{1}{2}x + 7$

50. $y = 3x + 2$ through $(6, 2)$

We have $m = 3$, $x_1 = 6$, and $y_1 = 2$. The new slope is $m_1 = -\dfrac{1}{3}$.

The equation of the line is

$y - 2 = -\dfrac{1}{3}(x - 6)$

$y - 2 = -\dfrac{1}{3}x + 2$

$y = -\dfrac{1}{3}x + 4$

3.2 Exercises

For exercises 1-10, find the x and y intercepts of the quadratic function. Then create a table of values and graph.

In order to find the x intercepts of any function, first set the function equal to zero. Then solve for x. In order to find the y intercepts of any function, set $x = 0$ and calculate $f(0)$. Once we have the x and y intercepts of a quadratic function, we can usually use two more points in order to create a graph.

2. $f(x) = x^2 + 2x + 1$

   For the x intercept(s), set the function equal to zero and solve:

   $x^2 + 2x + 1 = 0$

   $(x + 1)^2 = 0$

   $x = -1, -1$

   Therefore this function has only one x-intercept. To get the y intercept, set $x = 0$:

   $f(0) = 0^2 + 2(0) + 1 = 1$

   Therefore the y intercept of this function has coordinates $(0, 1)$. Next we will choose

   two values below $x = -1$ and above $x = 0$ to make the table of values and graph:

x	y
-3	4
-2	1
-1	0
0	1
1	4
2	9

   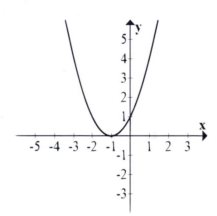

4. $g(x) = x^2 + 8x + 12$

   For the x intercept(s), set the function equal to zero and solve:

   $x^2 + 8x + 12 = 0$

   $(x + 6)(x + 2) = 0$

   $x = -6, -2$

   Therefore this function has two x-intercepts. To get the y intercept, set $x = 0$:

   $f(0) = 0^2 + 8(0) + 12 = 12$

   Therefore the y intercept of this function has coordinates $(0, 12)$. Next we will choose

   some values between $x = -6$ and $x = -2$ to make the table of values and graph:

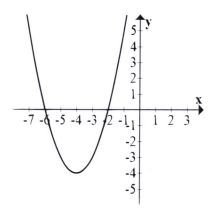

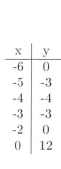

x	y
-6	0
-5	-3
-4	-4
-3	-3
-2	0
0	12

10. $f(x) = 3x^2 - 2x - 1$

For the x intercept(s), set the function equal to zero and solve:

$3x^2 - 2x - 1 = 0$

$(3x + 1)(x - 1) = 0$

$x = -\dfrac{1}{3},\ 1$

Therefore this function has two x-intercepts. To get the y intercept, set $x = 0$:

$f(0) = 3(0^2) - 2(0) - 1 = -1$

Therefore the y intercept of this function has coordinates $(0, -1)$. Next we will choose some values around $x = -\dfrac{1}{3}$ and $x = 1$ to make the table of values and graph:

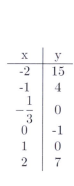

x	y
-2	15
-1	4
$-\dfrac{1}{3}$	0
0	-1
1	0
2	7

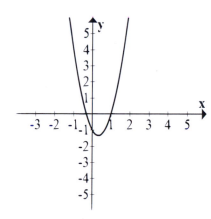

For exercises 11-20, use the method of completing the square to put each quadratic function into standard form. Identify the vertex, intercepts, and draw a graph of the parabola.

In order to put a quadratic function into standard form, replace $f(x)$ with y then subtract the constant term from both sides. If the coefficient of x^2 is not 1 then factor this from the right hand side. Then take half the coefficient of x and square it. Add this inside the parentheses (if they exist) on the right hand side. Multiply this value by the coefficient of x^2, then add the product to the left hand side. Factor the right hand side, then solve for y. Finally, replace y with $f(x)$. The quadratic function will be in standard form: $f(x) = a(x - h)^2 + k$. The location of the vertex has coordinates (h, k).

14. $g(x) = x^2 + 8x + 4$

Following the steps from above:

$$y = x^2 + 8x + 4$$

$$y - 4 = x^2 + 8x$$

$$y - 4 + 16 = x^2 + 8x + 16$$

$$y + 12 = (x + 4)^2$$

$$y = (x + 4)^2 - 12$$

$$f(x) = (x + 4)^2 - 12$$

The vertex is located at $(-4, -12)$. To find the x-intercepts we set $y = 0$ and solve for x:

$$(x + 4)^2 - 12 = 0$$

$$(x + 4)^2 = 12$$

$$x + 4 = \pm\sqrt{12}$$

$$x = -4 \pm 2\sqrt{3} \approx -7.46, \ -0.54$$

The y-intercept is located at $f(0) = 4$. A graph appears below:

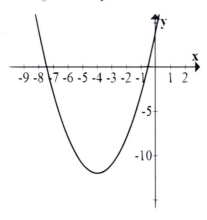

20. $f(x) = 3x^2 - 2x + 1$

Following the steps from above:

$$y = 3x^2 - 2x + 1$$

$$y - 1 = 3x^2 - 2x$$

$$y - 1 = 3\left(x^2 - \frac{2}{3}x\right)$$

$$y - 1 + \frac{1}{3} = 3\left(x^2 - \frac{2}{3}x + \frac{1}{9}\right)$$

$$y - \frac{2}{3} = 3\left(x - \frac{1}{3}\right)^2$$

$$y = 3\left(x - \frac{1}{3}\right)^2 + \frac{2}{3}$$

$$f(x) = 3\left(x - \frac{1}{3}\right)^2 + \frac{2}{3}$$

The vertex is located at $\left(\frac{1}{3}, \frac{2}{3}\right)$. To find the

x-intercepts we set $y = 0$

and solve for x:

$$3\left(x - \frac{1}{3}\right)^2 + \frac{2}{3} = 0$$

$$3\left(x - \frac{1}{3}\right)^2 = -\frac{2}{3}$$

This equation has no real solutions, because the right hand side is negative. Therefore this

graph has no x-intercepts.

The y-intercept is located at $f(0) = 1$. A graph appears below:

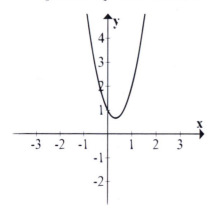

For exercises 21-30, find the vertex of each parabola, put $f(x)$ into standard form, and create a graph.

For these problems we will use the formula for the x-coordinate of the vertex: $x = -\dfrac{b}{2a}$. Remember that a is the coefficient of x^2 and b is the coefficient of x. To get the y-coordinate of the vertex, we will substitute the x-coordinate into the function. The final answer is of the form $f(x) = a(x-h)^2 + k$, where h is the x-coordinate of the vertex, and k is the y-coordinate.

22. $g(x) = x^2 - 4x + 5$

We have $a = 1$ and $b = -4$. Therefore the x-coordinate of the vertex is

$$x = -\frac{b}{2a} = -\frac{-4}{2(1)} = 2.$$ The y-coordinate of

the vertex is $f(2) = -3$. Therefore the standard form of the parabola is

$g(x) = (x-2)^2 - 3$. A graph of the parabola appears below:

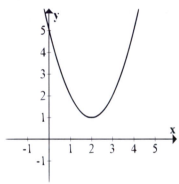

28. $p(x) = 3x^2 - 18x + 9$

We have $a = 3$ and $b = -18$. Therefore the x-coordinate of the vertex is

$$x = -\frac{b}{2a} = -\frac{-18}{2(3)} = 3.$$ The y-coordinate of the vertex is $f(3) = -18$.

Therefore the standard form of the parabola is $p(x) = 3(x-3)^2 - 18$.

A graph of the parabola appears below:

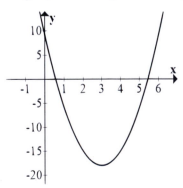

3.3   Exercises

In exercises 1 - 10, identify the degree of each polynomial function. Then determine the leading coefficient, number of terms, and constant term.

Remember that the degree of a polynomial is equal to the highest exponent that appears in the polynomial. The leading coefficient is the coefficient of the term with the highest degree. A term is of the form ax^n, where a is a real number and n is an integer greater than or equal to zero. The constant term is the only term without a variable. If there is no term without a variable then the constant term is zero.

2. $q(x) = 2x^3 - 3x + 5$

This polynomial has three terms. The degree is 3 and the leading coefficient is 2.

The constant term is 5.

4. $p(x) = 4x^5 + 2x^4 - 3x^3 + x^2 - x$

This polynomial has 5 terms. The degree is 5 and the leading coefficient is 4. The

constant term is 0.

8. $g(x) = \frac{2}{3}x^3 - \frac{4}{3}x^2 + x - 4$

This polynomial has 4 terms. The degree is 3 and the leading coefficient is $\frac{2}{3}$. The

constant term is -4.

In exercises 11 - 20, determine the zeroes of the given polynomial function. Then write the function in factored form.

Given a polynomial $p(x)$, a zero of the polynomial is any solution of the equation $p(x) = 0$. Every real zero of a polynomial corresponds to an x-intercept of its graph.

12. $q(x) = x^2 - 2x - 8$

To find the zeroes of the polynomial we set it equal to zero and solve:

$x^2 - 2x - 8 = 0$

$(x + 2)(x - 4) = 0$

The zeroes of this polynomial are $x = -2, 4$. The factored form is

$q(x) = (x + 2)(x - 4)$.

16. $q(x) = x^3 - 2x^2 - 15x$

   To find the zeroes of the polynomial we set it equal to zero and solve:

   $x^3 - 2x^2 - 15x = 0$

   $x(x^2 - 2x - 15) = 0$

   $x(x + 3)(x - 5) = 0$

   The zeroes of this polynomial are $x = -3, 0, 5$. The factored form is

   $q(x) = x(x + 3)(x - 5)$.

20. $m(n) = 3n^3 - 2n^2 - 12n + 8$

   We set the polynomial equal to zero. This time we need synthetic division:

   $3n^3 - 2n^2 - 12n + 8 = 0$

   We will try $n = 2$:

   ```
   2 | 3   -2   -12    8
     |      6     8   -8
     -------------------
       3    4    -4 |  0
   ```

   Since we got a zero in the box, we know that $n = 2$ is a zero. The new equation is

   $(n - 2)(3n^2 + 4n - 4) = 0$

   $(n - 2)(3n - 2)(n + 2) = 0$

   The zeroes of this function are $n = -2, \dfrac{2}{3}, 2$. The factored form is

   $m(n) = (n - 2)(3n - 2)(n + 2)$.

In exercises 21 - 30, create a table of values and graph the function. Be sure that the table of values contains the roots of the polynomial function (Note: these are the same functions from problems 11 - 20).

22. $q(x) = x^2 - 2x - 8$

   We know the zeroes of this function are $x = -2, 4$. We will choose some other x values

   near these, as well as $x = 0$:

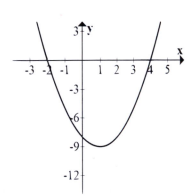

x	y
-3	7
-2	0
-1	-5
0	-8
1	-9
2	-8

26. $q(x) = x^3 - 2x^2 - 15x$

We know the zeroes of this function are $x = -3, 0, 5$. We will choose some other x

values near these to complete the table of values:

x	y
-4	-36
-3	0
-2	14
-1	12
0	0
1	-16
2	-30
3	-36
4	-28
5	0

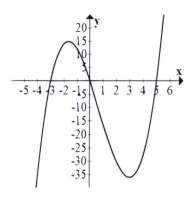

30. $m(n) = 3n^3 - 2n^2 - 12n + 8$

We know the zeroes of this function are $x = -2, 4$. We will choose some other x values

near these, as well as $x = 0$:

x	y
-2	15
-1	4
$-\dfrac{1}{3}$	0
0	-1
1	0
2	7

 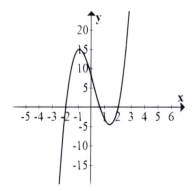

3.4   Exercises

Determine the maximum possible number of solutions for each equation.

The maximum possible number of solutions to a polynomial equation is equal to the degree
of the equation, which is the highest exponent that occurs.

2. $4x^3 + 2x = 5x^4 + 8x^2 + 4$

    The degree of this equation is 4, so that is the maximum possible number of solutions.

4. $4x^6 + 5x^4 - 3x^2 = 6$

    The degree of this equation is 6, so that is the maximum possible number of solutions.

6. $6x^3 + 4x^2 + 3x = x^5 + 2$

    The degree of this equation is 5, so that is the maximum possible number of solutions.

Given a function and one of its zeroes, find all of the zeroes of the function.

For these problems we will use synthetic division. This will help us in factoring the function. Remember that if we know that $x = a$ is a zero of a polynomial function $p(x)$, then $x - a$ is a factor.

8. $f(x) = x^3 - 4x^2 + 6x - 4;\ x = 2$

    We use synthetic division with $x = 2$:

    ```
    2 | 1  -4   6   -4
      |      2  -4    4
      -----------------
        1  -2   2 |  0
    ```

    We got zero in the box, as expected. The factored form of the polynomial is:

    $f(x) = (x - 2)(x^2 - 2x + 2)$

    The second factor does not factor. To get the other two zeroes we will set this equal to zero and use the quadratic formula:

    $x^2 - 2x + 2 = 0$

    $$x = \frac{-(-2) \pm \sqrt{(-2)^2 - 4(1)(2)}}{2(1)}$$

    $$= \frac{2 \pm \sqrt{4 - 8}}{2}$$

    $$= \frac{2 \pm \sqrt{-4}}{2}$$

    $$= \frac{2 \pm 2i}{2}$$

    $$= 1 \pm i$$

    Therefore the other two zeroes are $x = 1 - i$ and $x = 1 + i$.

12. $q(x) = x^4 - 3x^2 - 4; \; x = i$

We can factor this function:

$$q(x) = (x^2 - 4)(x^2 + 1) = (x - 2)(x + 2)(x^2 + 1)$$

Setting each of these factors equal to zero, we get the zeroes $x = -2, \, 2, \, \pm i$.

State the number of positive real zeros, negative real zeros, and imaginary zeros for each function.

We will use Descartes' Rule of signs for these problems.

14. $g(x) = x^4 + x^3 + x^2 - 2x - 1$

The signs of the coefficients go positive, positive, positive, negative, negative. They change exactly once (from positive 1 to negative 2). Therefore Descartes' Rule of Signs tells us that there must be exactly one positive zero. Next we calculate $g(-x) = (-x)^4 + (-x)^3 + (-x)^2 - 2(-x) - 1 = x^4 - x^3 + x^2 + 2x - 1$. The signs of the coefficients are positive, negative, positive, positive, negative, so they change exactly three times. Therefore there are either 3 negative zeroes, or 1 negative zero. If there are 3 negative zeroes then this accounts for all four zeroes (remember the one positive zero). If there is one negative zero then we must have 2 complex zeroes.

18. $f(x) = x^8 - x^5 + x^3 + x - 3$

The signs of the coefficients of $f(x)$ go positive, negative, positive, positive, negative. They change three times. So there must be either 3 positive zeroes or 1 positive zero. Now we calculate $f(-x) = (-x)^8 - (-x)^5 + (-x)^3 + (-x) - 3 = x^8 + x^5 - x^3 - x - 3$. The signs change exactly once, so there must be one negative zero. From the first part, if we have 3 positive zeroes then there must be 4 complex zeroes (this will add up to 8). If there is 1 positive zero then we must have 6 complex zeroes, 1 negative zero, and one positive zero.

Find a polynomial function of least degree with integer coefficients that has the given zeros.

For these problems we will use two facts. First, if $x = a$ is a zero of a polynomial, then $x - a$ must be a factor. Second, if $x = a + bi$ is a zero of a polynomial with real coefficients, then $x = a - bi$ must be a zero as well (Conjugate Roots Theorem).

20. 3, −2

We have the zeroes $x = 3$ and $x = -2$. Therefore the factors are $(x - 3)$ and $(x + 2)$. This gives

$$
\begin{aligned}
p(x) &= (x - 3)(x + 2) \\
&= x^2 + 2x - 3x - 6 \\
&= x^2 - x - 6
\end{aligned}
$$

24. $2i$, 4

We have $x = 2i$ and $x = 4$ as zeroes. The complex conjugate of $2i$ is $-2i$. The factors of our polynomial will be $(x - 4)$, $(x - 2i)$, and $(x + 2i)$. This gives

$$
\begin{aligned}
p(x) &= (x - 4)(x - 2i)(x + 2i) \\
&= (x - 4)(x^2 + 2ix - 2ix - 4i^2) \\
&= (x - 4)(x^2 + 4) \\
&= x^3 - 4x^2 + 4x - 16
\end{aligned}
$$

28. 5, $4 - i$

We have $x = 5$ and $x = 4 - i$. The conjugate of $4 - i$ is $4 + i$. The factors of our polynomial are $(x - 5)$, $(x - (4 - i))$, and $(x - (4 + i))$. This gives

$$
\begin{aligned}
p(x) &= (x - 5)(x - (4 - i))(x - (4 + i)) \\
&= (x - 5)(x - 4 + i)(x - 4 - i) \\
&= (x - 5)((x - 4) + i)((x - 4) - i) \\
&= (x - 5)((x - 4)^2 - i^2) \\
&= (x - 5)(x^2 - 8x + 16 - (-1)) \\
&= (x - 5)(x^2 - 8x + 17) \\
&= x^3 - 8x^2 + 17x - 5x^2 + 40x - 85 \\
&= x^3 - 13x^2 + 57x - 85
\end{aligned}
$$

3.5 Exercises

For each of the following functions, calculate any asymptotes (vertical, horizontal, or oblique), create a table of values, and graph the function.

We are looking for vertical, horizontal, and oblique asymptotes. x-intercepts occur when the numerator of the function equals zero. y-intercepts occur when $x = 0$.

4. $g(x) = \dfrac{3}{x - 2}$

To find vertical asymptotes, we set the denominator equal to zero. This gives $x = 2$. Therefore $x = 2$ is a vertical asymptote.

For the horizontal asymptote we use the fact that the degree of the numerator is 0 and the degree of the denominator is 1. Since the degree of the numerator is less than that of the denominator, $y = 0$ is a horizontal asymptote.

Since the numerator is constant, it can never equal zero. Therefore the graph has no x-intercepts.

The y-intercept occurs when $x = 0$. This gives $g(0) = \dfrac{3}{0 - 2} = -\dfrac{3}{2}$.

A table of values and graph appears below:

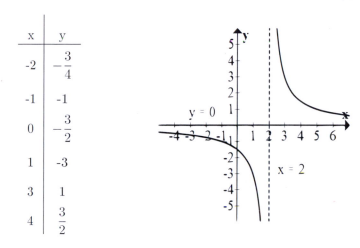

x	y
-2	$-\dfrac{3}{4}$
-1	-1
0	$-\dfrac{3}{2}$
1	-3
3	1
4	$\dfrac{3}{2}$

8. $h(x) = \dfrac{x^2 - 2x - 8}{x^2 - 9}$

To find vertical asymptotes, we set the denominator equal to zero. This gives

$$
\begin{aligned}
x^2 - 9 &= 0 \\
(x + 3)(x - 3) &= 0 \\
x &= -3,\ 3
\end{aligned}
$$

Therefore $x = -3$ and $x = 3$ are both vertical asymptotes.

For the horizontal asymptote we use the fact that the degree of the numerator is 2 and the degree of the denominator is 2. Since the degrees are equal, we take the ratio of the leading coefficients. Therefore $y = 1$ is a horizontal asymptote.

To find the x-intercepts we set the numerator equal to zero:

$x^2 - 2x - 8 = 0$

$(x + 2)(x - 4) = 0$

Therefore $x = -2$ and $x = 4$ are the x-intercepts.

The y-intercept occurs when $x = 0$. This gives $h(0) = \dfrac{-8}{-9} = \dfrac{8}{9}$.

A table of values and graph appears below:

x	y
-4	$\frac{16}{7}$
-2	0
-1	$\frac{5}{8}$
0	$\frac{8}{9}$
1	$\frac{9}{8}$
2	$\frac{8}{5}$
5	$\frac{7}{16}$

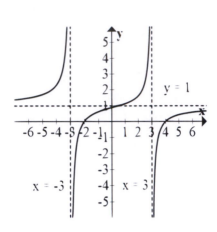

12. $f(x) = \dfrac{x^3 - 9x}{x^2 - 4}$

To find vertical asymptotes, we set the denominator equal to zero. This gives

$x^2 - 4 = 0$

$(x + 2)(x - 2) = 0$

$x = \pm 2$

Therefore $x = -2$ and $x = 2$ are the vertical asymptotes.

The degree of the numerator is 3 and the degree of the denominator is 2. Since the degree of the numerator is one more than that of the denominator, we have an oblique asymptote. Using long division, we get $f(x) = x - \dfrac{5x}{x^2 - 4}$, so the equation of the oblique asymptote is $y = x$.

To get the x-intercepts we set the numerator equal to zero:

$x^3 - 9x = 0$
$x(x^2 - 9) = 0$
$x(x + 3)(x - 3) = 0$
The x-intercepts are $x = -3$, $x = 0$, and $x = 3$.

The y-intercept occurs when $x = 0$. This gives $g(0) = \dfrac{0}{0 - 4} = 0$.

A table of values and graph appears below:

x	y
-4	$-\dfrac{7}{3}$
-3	0
-1	$-\dfrac{8}{3}$
0	0
1	3
3	0
4	$\dfrac{7}{3}$

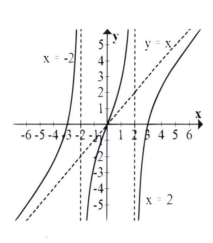

3.6   Exercises

Find the unique lowest degree polynomial passing through each set of points. Then graph the polynomial and verify the points lie on the graph.

Recall that if we have two points, this determines a straight line. With three noncollinear points, a parabola passes through (degree 2 polynomial). With four points we have a cubic graph, etc.

2. We are given the points $\{(-3, 1), (-1, 5)\}$.

We have two points, so we will have a straight line. First we will calculate the slope, then use the point-slope form.

We have $x_1 = -3$, $y_1 = 1$, $x_2 = -1$, and $y_2 = 5$. Therefore

$$m = \frac{y_2 - y_1}{x_2 - x_1} = \frac{5 - 1}{(-1) - (-3)} = \frac{4}{-1 + 3} = \frac{4}{2} = 2$$

Using the point-slope form:

$$\begin{aligned} y - 1 &= 2(x - (-3)) \\ y - 1 &= 2(x + 3) \\ y - 1 &= 2x + 6 \\ y &= 2x + 7 \end{aligned}$$

Therefore our desired polynomial is $p(x) = 2x + 7$.

6. We are given the points $\{(-2, 10), (-1, 0), (3, 20)\}$.

We have three noncollinear points (they don't lie on a straight line). Therefore a parabola passes through them. The general function for a parabola is $p(x) = ax^2 + bx + c$. We need to find a, b, and c. We substitute each point into this equation:

For the point $(-2, 10)$ we get $10 = a(-2)^2 + b(-2) + c$ so $4a - 2b + c = 10$.

For the point $(-1, 0)$ we get $0 = a(-1)^2 + b(-1) + c$ so $a - b + c = 0$.

For the point $(3, 20)$ we get $20 = a(3^2) + b(3) + c$ so $9a + 3b + c = 20$.

This gives the system of equations

$$\begin{cases} 4a - 2b + c &= 10 \\ a - b + c &= 0 \\ 9a + 3b + c &= 20 \end{cases}$$

If we subtract the first equation from the third equation, and then the second equation from the third equation, we get the system

$$\begin{cases} 5a + 5b &= 10 \\ 8a + 4b &= 20 \end{cases}$$

Now we divide the first equation by 5 and the second equation by 4, getting the following system:

$$\begin{cases} a + b &= 2 \\ 2a + b &= 5 \end{cases}$$

Subtracting the first equation from the second equation gives $a = 3$. Next substitute $a = 3$ into the equation $a + b = 2$. This gives $b = -1$. Substituting these both into the original second equation, we get $3 - (-1) + c = 0$, so $c = -4$.

Now we replace the values of a, b, and c into the function $p(x) = ax^2 + bx + c$, getting $p(x) = 3x^2 - x - 4$.

8. We are given the points $\{(-2, 12), (-1, 12), (1, 0), (2, 0)\}$.

We have four points, so we will look for a cubic function. The general cubic function is $p(x) = ax^3 + bx^2 + cx + d$.

The first point is $(-2, 12)$, which gives $12 = a(-2)^3 + b(-2)^2 + c(-2) + d$, so $-8a + 4b - 2c + d = 12$.

The second point is $(-1, 12)$, so $12 = a(-1)^3 + b(-1)^2 + c(-1) + d$, so $-a + b - c + d = 12$.

The third point is $(1, 0)$, so $0 = a(1^3) + b(1^2) + c(1) + d$ and $a + b + c + d = 0$

The fourth point is $(2, 0)$, so $0 = a(2^3) + b(2^2) + c(2) + d$ and $8a + 4b + 2c + d = 0$.

Combining these four equations leads to the system

$$\begin{cases} -8a + 4b - 2c + d &= 12 \\ -a + b - c + d &= 12 \\ a + b + c + d &= 0 \\ 8a + 4b + 2c + d &= 0 \end{cases}$$

We will subtract each of the first three equations from the last equation. This gives the system

$$\begin{cases} 16a + 4c & = & -12 \\ 9a + 3b + 3c & = & -12 \\ 7a + 3b + c & = & 0 \end{cases}$$

Now we divide the first equation by 4 and the second equation by 3:

$$\begin{cases} 4a + c & = & -3 \\ 3a + b + c & = & -4 \\ 7a + 3b + c & = & 0 \end{cases}$$

Now we subtract the first two equations from the third equation:

$$\begin{cases} 3a + 3b & = & 3 \\ 4a + 2b & = & 4 \end{cases}$$

Divide the first equation by 3 and the second equation by 2:

$$\begin{cases} a + b & = & 1 \\ 2a + b & = & 2 \end{cases}$$

Subtract the first equation from the second equation. This gives $a = 1$. Substitute this into the first equation. This gives $b = 0$. Substitute these into the first equation from the previous system. This gives $c = -7$. Substitute these into any of the first equations. This gives $d = 6$.

The cubic polynomial is $p(x) = x^3 - 7x + 6$.

Chapter 4

Exponential and Logarithmic Functions

4.1   Exercises

For each of the functions in problems 1 - 10, evaluate $f(-2)$, $f(0)$ and $f(2)$.

2. $f(x) = 5^x$

$$f(-2) = 5^{-2} = \frac{1}{25}$$

$$f(0) = 5^0 = 1$$

$$f(2) = 5^2 = 25$$

4. $f(x) = \left(\frac{2}{3}\right)^x$

$$f(-2) = \left(\frac{2}{3}\right)^{-2} = \frac{9}{4}$$

$$f(0) = \left(\frac{2}{3}\right)^0 = 1$$

$$f(2) = \left(\frac{2}{3}\right)^2 = \frac{4}{9}$$

6. $f(x) = 3^{x+1}$

$$f(-2) = 3^{-2+1} = \frac{1}{3}$$

$$f(0) = 3^{0+1} = 3$$

$$f(2) = 3^{2+1} = 27$$

10. $f(x) = (x-1) \cdot 3^x$

$$f(-2) = (-2-1) \cdot 3^{-2} = -\frac{1}{3}$$

$$f(0) = (0-1) \cdot 3^0 = -1$$

$$f(2) = (2-1) \cdot 3^2 = 9$$

Create a table of values and graph each of the functions in problems 11 - 20. Use the x values *{-2, -1, 0, 1, 2}.*

14. $g(x) = \left(\dfrac{1}{2}\right)^x$

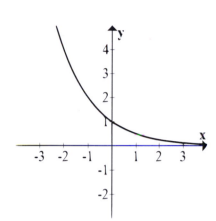

x	y
-2	4
-1	2
0	1
1	$\dfrac{1}{2}$
2	$\dfrac{1}{4}$

20. $g(x) = 4^{3-2x} - 1$

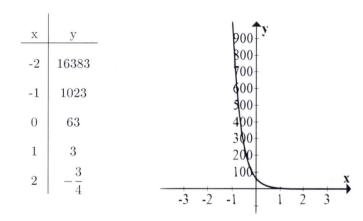

x	y
-2	16383
-1	1023
0	63
1	3
2	$-\dfrac{3}{4}$

24. The pressure in a tire with a leak is initially 32 pounds per square inch (psi) and can be modeled by $p(t) = 32(0.95)^t$ after t minutes.

a. What is the pressure in the tire after 10 minutes?

After 10 minutes, the pressure will be $p(10) = 32(0.95)^{10} \approx 19.16$ psi.

b. When will the pressure reach 20 psi?

The pressure will reach 20 psi when $p(t) = 20$. This leads to the equation $32(0.95)^t = 20$. If we divide both sides by 32 then we get $(0.95)^t = 0.625$. In order to find the exact answer, we need logarithms. We can use trial and error. After 9 minutes the pressure is approximately 20.168 psi, and after 10 minutes the pressure is approximately 19.16 psi.

4.2 Exercises

In problems 1 - 10, calculate the value of each logarithm by converting to an equivalent exponential expression.

For the logarithmic expression $y = \log_a x$, we assign each variable as follows: $a =$ base, $y =$ exponent, $x =$ result. The equivalent exponential expression is then $a^y = x$.

2. $y = \log_9 243$

   base $= 9$, exponent $= y$, result $= 243$

   $9^y = 243$

   $y = \dfrac{5}{3}$

4. $y = \log_{\frac{2}{3}} 3.375$

   base $= \dfrac{2}{3}$, exponent $= y$

   result $= 3.375 = \dfrac{27}{8}$

   $\left(\dfrac{2}{3}\right)^y = \dfrac{27}{8}$

   $y = -3$

6. $y = \log_{16} 0.25$

   base $= 16$, exponent $= y$, result $= 0.25 = \dfrac{1}{4}$

   $16^y = \dfrac{1}{4}$

   $y = -\dfrac{1}{2}$

10. $y = \log_{1.2} 1.728$

    base $= 1.2$, exponent $= y$

    result $= 1.728 = 1.2^3$

    $1.2^y = 1.728$

    $y = 3$

For problems 11 - 20, evaluate $f(-2)$, $f(0)$ and $f(2)$.

In these problems remember that $\log x$ is base 10, and $\ln x$ is base e. Furthermore, we cannot take the log of a negative number, regardless of base.

12. $f(x) = \ln x$

    $f(-2) = \ln(-2)$, which does not exist

    $f(0) = \ln 0$, which does not exist

    $f(2) = \ln 2 \approx 0.6931$

14. $f(x) = \ln(x - 2)$

    $f(-2) = \ln(-4)$, which does not exist

    $f(0) = \ln(0 - 2) = \ln(-2)$, which does not exist

    $f(2) = \ln(2 - 2) = \ln 0$, which does not exist

18. $f(x) = 3 \log x + 2$

    $f(-2) = 3 \log(-2) + 2$, which does not exist

    $f(0) = 3 \log 0 + 2$, which does not exist

    $f(2) = 3 \log 2 + 2 \approx 3.903$

20. $f(x) = 3 \log(x + 2)$

    $f(-2) = 3 \log(-2 + 2)$, which does not exist

    $f(0) = 3 \log(0 + 2) \approx 0.903$

    $f(2) = 3 \log(2 + 2) = 3 \log 4 \approx 1.806$

For problems 21 - 30, create a table of values and graph for each function.

22. $f(x) = \ln x$

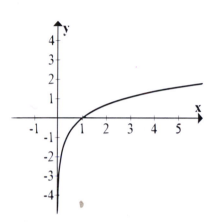

x	y
0.1	-2.303
0.5	-0.693
1	2
2	0.693
3	1.097

24. $f(x) = \ln(x - 2)$

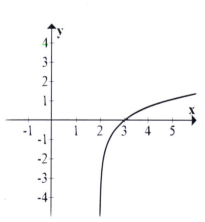

x	y
2.1	-2.303
2.5	-0.693
3	2
4	0.693
5	1.097

32. The relationship between pH and the hydrogen ion concentration is given by the formula $-\log[H+] = pH$, where $[H+]$ is the symbol for hydrogen ion concentration.

a) What is the pH of water if it has a $[H+]$ concentration of $1.0 \cdot 10^{-7}$ M?

We are given $[H+] = 1.0 \cdot 10^{-7}$. Therefore $pH = -\log 1.0 \cdot 10^{-7} = -\log 10^{-7}$. If we multiply both sides by -1 then we get $-pH = \log 10^{-7}$. The base is 10, the exponent is $-pH$, and the result is 10^{-7}. This gives $10^{-pH} = 10^{-7}$, so $pH = 7$.

b) What is the [H+] concentration of a substance whose pH is 6? (Hint: first divide by -1 to get rid of the negative sign on the right hand side of the formula, then convert to exponent form to solve for pH).

We are given $pH = 6$, so $-\log[H+] = 6$. If we divide both sides by -1 then we get $\log[H+] = -6$. The base is 10, the exponent is -6, and the result is [H+]. Therefore [H+] $= 10^{-6}$.

4.3 Exercises

In problems 1 - 10, calculate the value of each logarithm using the change of base formula and your calculator.

The change of base formula tells us that $\log_a b = \dfrac{\log_c b}{\log_c a}$ for any positive bases a, b, c not equal to 1. If $c = 10$ then $\log_a b = \dfrac{\log b}{\log a}$, and we can use our calculator to get an approximation to the answer.

2. $\log_6 9 = \dfrac{\log 9}{\log 6} \approx 1.226$

4. $\log_{\frac{2}{3}} 5 = \dfrac{\log 5}{\log \frac{2}{3}} \approx -3.969$

6. $\log_3(0.2) = \dfrac{\log 0.2}{\log 3} \approx -1.465$

10. $\log_{1.2} 22 = \dfrac{\log 22}{\log 1.2} \approx 16.954$

In exercises 11 - 30, use the properties of logarithms to expand the expression as the sum, difference, and/or multiple of logarithms.

For the following problems we will use these properties of logarithms:

Exponent Rule : $\log_a x^r = r\log_a x$ for any real number r

Product Rule : $\log_a(x \cdot y) = \log_a x + \log_a y$

Quotient Rule : $\log_a \dfrac{x}{y} = \log_a x - \log_a y$

12. $\log_8 3y = \log_8 3 + \log_8 y$ (Product Rule)

14. $\log_5 \dfrac{z}{2} = \log_5 z - \log_5 2$ (Quotient Rule)

16. $\log_{11}(x-2)(x+3) = \log_{11}(x-2) - \log_{11}(x+3)$ (Quotient Rule)

18. $\log_{13} y(y-2)(z+1)^2 = \log_{13} y + \log_{13}(y-2) + \log_{13}(z+1)^2$

$\quad\quad\quad\quad = \log_{13} y + \log_{13}(y-2) + 2\log_{13}(z+1)$ (Product Rule then Exponent Rule)

22. $\ln \dfrac{x(x+1)}{x+2} = \ln(x(x+1)) - \ln(x+2) = \ln x + \ln(x+1) - \ln(x+2)$

      (Quotient Rule then Product Rule)

24. $\ln \sqrt{x(x+2)} = \ln\left(x(x+2)\right)^{1/2} = \dfrac{1}{2}\ln(x(x+2)) = \dfrac{1}{2}(\ln x + \ln(x+2))$

      (Exponent Rule then Product Rule)

26. $\ln \dfrac{x^2\sqrt{y}}{z^3} = \ln\left(x^2 y^{1/2}\right) - \ln z^3 = \ln x^2 + \ln y^{1/2} - \ln z^3 = 2\ln x + \dfrac{1}{2}\ln y - 3\ln z$

      (Quotient Rule, then Product Rule, then Exponent Rule)

In exercises 31 - 50, write each expression as a single logarithm.

Again, for these problems, we will use a combination of the Product, Quotient, and Exponent Rules for logarithms.

32. $\log(x-2) + \log x = \log x(x-2)$ (Product Rule)

34. $\log(x-2) - \log x = \log \dfrac{x-2}{x}$ (Quotient Rule)

36. $3\log_2(x-1) = \log_2(x-1)^3$ (Exponent Rule)

38. $3\log(x-1) + 2\log(x+1) = \log(x-1)^3 + \log(x+1)^2 = \log(x-1)^3(x+1)^2$

      (Exponent Rule then Product Rule)

40. $\dfrac{1}{3}\log(x+3) = \log(x+3)^{1/3} = \log \sqrt[3]{x+3}$ (Exponent Rule)

42. $\ln(x+2) - 3\ln(x-2) = \ln(x+2) - \ln(x-2)^3 = \ln \dfrac{x+2}{(x-2)^3}$

      (Exponent Rule then Quotient Rule)

48. $\log_2 x - \dfrac{1}{3}[2\log_2(x-2) + 3\log_2(x+1)] = \log_2 x - \dfrac{1}{3}\left[\log_2(x-2)^2 + \log_2(x+1)^3\right]$

      $= \log_2 x - \log_2\left((x-2)^2(x+1)^3\right)^{1/3} = \log_2 \dfrac{x}{\log_2\left((x-2)^2(x+1)^3\right)^{1/3}}$

      (Exponent Rule then Product Rule then Quotient Rule)

Solve for x.

For these problems we will get a single logarithm on one side of the equation using the properties of logs. Then convert the equation into an exponential equation and solve the resulting equation. It is necessary to check the answer in order to ensure we are not taking the log of a negative number or zero.

52.

$$
\begin{aligned}
\log_4(x+6) - \log_4(x-2) &= 2 \\
\log 4 \frac{x+6}{x-2} &= 2 \\
\frac{x+6}{x-2} &= 4^2 \\
\frac{x+6}{x-2} &= 16 \\
x+6 &= 16(x-2) \\
x+6 &= 16x - 32 \\
15x &= 38 \\
x &= \frac{38}{15}
\end{aligned}
$$

54.

$$
\begin{aligned}
\ln(x+1) - \ln(x-2) &= \ln x \\
\ln \frac{x+1}{x-2} &= \ln x \\
\frac{x+1}{x-2} &= x \\
x+1 &= x(x-2) \\
x^2 - 2x &= x+1 \\
x^2 - 3x - 1 &= 0
\end{aligned}
$$

This equation does not factor. We will use the quadratic formula with $a = 1$, $b = -3$, and $c = -1$:

$$
\begin{aligned}
x &= \frac{-(-3) \pm \sqrt{(-3)^2 - 4(1)(-1)}}{2(1)} \\
&= \frac{3 \pm \sqrt{9+4}}{2} \\
&= \frac{3 \pm \sqrt{13}}{2}
\end{aligned}
$$

$x = \dfrac{3 - \sqrt{13}}{2}$ is negative, so when we substitute this into the original equation, the argument of the second logarithm is negative. Therefore this does not solve the equation. However

$x = \dfrac{3 + \sqrt{13}}{2}$ does.

56.

$$\begin{aligned}
\log_3 x + \log_3(x - 8) &= 2 \\
\log_3 x(x - 8) &= 2 \\
x(x - 8) &= 3^2 \\
x^2 - 8x - 9 &= 0 \\
(x + 1)(x - 9) &= 0
\end{aligned}$$

Therefore $x = -1, 9$. We cannot use a negative value in the first logarithm in the equation $(\log_3 x)$. However $x = 9$ does solve the equation.

60. $\log 4x - \log(12 + \sqrt{x}) = 2$

$$\begin{aligned}
\log 4x - \log(12 + \sqrt{x}) &= 2 \\
\log \frac{4x}{12 + \sqrt{x}} &= 2 \\
\frac{4x}{12 + \sqrt{x}} &= 10^2 \\
4x &= 100(12 + \sqrt{x}) \\
4x - 100\sqrt{x} - 1200 &= 0
\end{aligned}$$

This equation is quadratic in form. We will substitute $u = \sqrt{x}$, so $u^2 = x$. The equation becomes

$$\begin{aligned}
4u^2 - 100u - 1200 &= 0 \\
u^2 - 25u - 300 &= 0 \\
u &= \frac{-(-25) \pm \sqrt{(-25)^2 - 4(1)(-300)}}{2(1)} \\
&= \frac{25 \pm \sqrt{625 + 1200}}{2} \\
&= \frac{25 \pm \sqrt{1825}}{2} \\
&= \frac{25 \pm 5\sqrt{73}}{2}
\end{aligned}$$

Since $u = \sqrt{x}$, this means that $x = u^2$ and $x = \left(\dfrac{25 \pm 5\sqrt{73}}{2}\right)^2$.

4.4 Exercises

In problems 1 - 30, find the solution set for each equation.

In these problems we will make use of the fact that if $a^u = a^v$ for a positive $a \neq$, then $u = v$. So if we can write both sides of the equation with the same base, the exponents must be equal.

2. $3^x = 729$

$3^x = 3^6$

$x = 6$

4. $7^x = \dfrac{1}{49}$

$7^x = 7^{-2}$

$x == -2$

6. $\left(\dfrac{5}{2}\right)^x = \dfrac{16}{625}$

$\left(\dfrac{5}{2}\right)^x = \left(\dfrac{5}{2}\right)^{-4}$

$x = -4$

8. $5^x = 1$

$5^x = 5^0$

$x = 0$

10. $3^{3x-2} = 81$

$3^{3x-2} = 3^4$

$3x - 2 = 4$

$3x = 6$

$x = 2$

12. $5^{2x+5} = 125$

$5^{2x+5} = 5^3$

$2x + 5 = 3$

$2x = -2$

$x = -1$

18. $4^{x^2-x} = 16$

$4^{x^2-x} = 4^2$

$x^2 - x = 2$

$x^2 - x - 2 = 0$

$(x+2)(x-3) = 0$

$x = -2,\ x = 3$

20. $2^{x^2+5x} = 64$

$2^{x^2+5x} = 2^6$

$x^2 + 5x = 6$

$x^2 + 5x - 6 = 0$

$(x+1)(x-6) = 0$

$x = -1,\ x = 6$

22. $e^{2x} - e^x - 2 = 0$

Define $u = e^x$, so $u^2 = e^{2x}$

$u^2 - u - 2 = 0$

$(u+1)(u-2) = 0$

24. $4^{-2x} + 2 \cdot 4^{-x} - 8 = 0$

Define $u = 4^{-x}$, so $u^2 = 4^{-2x}$

$u^2 + 2u - 8 = 0$

$(u+4)(u-2) = 0$

$u = -1$ or $u = 2$ $u = -4$ or $u = 2$

$e^x = -1$ or $e^x = 2$ $4^{-x} = -4$ or $4^{-x} = 2$

Only the second equation has a solution   Only the second equation has a solution

$x = \ln 2$ $x = -\dfrac{1}{2}$

26. $6^x - 1 - 12 \cdot 6^{-x} = 0$ 28. $\dfrac{250}{2 - e^x} = 300$

$6^x \left(6^x - 1 - 12 \cdot 6^{-x} \right) = 0$ $250 = 300(2 - e^x)$

$6^{2x} - 6^x - 12 = 0$ $250 = 600 - 300e^x$

Define $u = 6^x$ so $u^2 = 6^{2x}$ $300e^x = 350$

$u^2 - u - 12 = 0$ $e^x = \dfrac{350}{300} = \dfrac{7}{6}$

$(u + 3)(u - 4) = 0$ $x = \ln\left(\dfrac{7}{6}\right)$

$u = -3$ or $u = 4$

$6^x = -3$ or $6^x = 4$

Only the second equation has a solution.

$x = \log_6 4 \approx 0.7737$

4.5 Exercises

In problems 1 - 8, calculate the simple interest earned for each scenario.

For these problems we use the simple interest formula:

$$I = Prt$$

In this formula P is the initial principle (amount borrowed or invested), r is the annual percentage rate expressed as a decimal, and t is the number of years.

2. $750 at 4% for 5 years

    We have $P = 750$, $r = 0.04$, $t = 5$:

    $I = Prt = (750)(0.04)(5) = \$150$

6. $5000 at 4.5% for 8 years

    We have $P = 5000$, $r = 0.045$, $t = 8$:

    $I = Prt = (5000)(0.045)(8) = \$1,800$

In problems 9 - 16, calculate the compound interest earned for each scenario.

The compound interest formula is

$$A = P\left(1 + \frac{r}{n}\right)^{nt}$$

Here A is the final amount of money (principle plus interest), P is the principle, r is the annual interest rate expressed as a decimal, n is the compounding frequency (number of times per year interest is compounded), and t is the number of years of the investment.

10. $2500 at 5% compounded quarterly for 6 years

   We have $P = 2500$, $r = 0.05$, $n = 4$, $t = 6$:

   $$A = P\left(1 + \frac{r}{n}\right)^{nt} = 2500\left(1 + \frac{0.05}{4}\right)^{4(6)} = 2500\,(1.0125)^{24} = \$3,368.38$$

14. $6000 at 4% compounded daily for 4 years

   We have $P = 6000$, $r = 0.04$, $n = 365$, $t = 4$:

   $$A = P\left(1 + \frac{r}{n}\right)^{nt} = 6000\left(1 + \frac{0.04}{365}\right)^{365(4)} = \$7041.00$$

In problems 17 - 20, calculate the APY.

The formula for APY is

$$APY = \left(1 + \frac{r}{n}\right)^{n} - 1$$

Here r is the annual percentage rate expressed a decimal and n is the compounding frequency. The final answer will be a percentage expressed in decimal form.

18. $2500 at 5% compounded quarterly for 6 years

   We have $r = 0.05$ and $n = 4$:

   $$APY = \left(1 + \frac{r}{n}\right)^{n} - 1 = \left(1 + \frac{0.05}{4}\right)^{4} - 1 = 1.0125^4 - 1 \approx 0.0509$$

   The annual percentage yield is approximately 5.09%.

20. $10000 at 5.6% compounded monthly for 10 years

   We have $r = 0.056$ and $n = 12$:

   $$APY = \left(1 + \frac{r}{n}\right)^{n} - 1 = \left(1 + \frac{0.056}{12}\right)^{12} - 1 \approx 0.0575$$

   The annual percentage yield is approximately 5.75%.

24. **Newtons Law of Cooling** Nayky has been pulling an all-nighter and is feeling some hunger pangs. He throws a pizza in the oven. Half an hour later, he takes the pizza out. The oven is at 350° and the room temperature is 72°. Then a formula for calculating the temperature of the pizza t minutes after removal from the oven is

$$T(t) = H_S + (H_0 - H_S)e^{-kt}$$

where H_S represents the air temperature, H_0 represents the oven temperature, and k is an unknown constant. We will measure time in minutes. In this problem we will use $k = .054$.

a.   Write out the function using the numbers presented in this problem.

We have $H_S = 72$, $H_0 = 350$, $k = 0.054$, $t = 30$:

$$T(t) = H_S + (H_0 - H_S)e^{-kt} = 72 + (350 - 72)e^{-0.054t} = 72 + 278e^{-0.054t}$$

b.   Create a graph of the function written in part a.

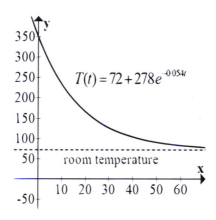

c.   What will the temperature of the pizza be after 30 minutes?

We substitute $t = 30$ into our formula:

$$T(30) = 72 + 278e^{-0.054t(30)} \approx 127^o \text{ Fahrenheit}$$

d.   How long will it take for the pizza to cool to a temperature of 250°?

We need to solve the equation $T(t) = 250$ for t:

$$
\begin{aligned}
72 + 278e^{-0.054t} &= 250 \\
278e^{-0.054t} &= 178 \\
e^{-0.054t} &= \frac{178}{278} \\
-0.054t &= \ln\left(\frac{178}{278}\right) \\
t &= -\frac{\ln\left(\frac{178}{278}\right)}{0.054} \\
&\approx 8.26
\end{aligned}
$$

So it will only take just over 8 minutes to cool to 250° Fahrenheit.

Chapter 5

Trigonometric Functions

5.1 Exercises

Convert each of the following degree measures into radians.

In order to convert a degree measure into radians, multiply by $\dfrac{\pi}{180}$.

2. $75^o \cdot \dfrac{\pi}{180} = \dfrac{75\pi}{180} = \dfrac{5\pi}{12}^R$

4. $150^o \cdot \dfrac{\pi}{180} = \dfrac{150\pi}{180} = \dfrac{5\pi}{6}^R$

10. $375^o \cdot \dfrac{\pi}{180} = \dfrac{375\pi}{180} = \dfrac{25\pi}{12}^R$

Convert each of the following radian measures into degrees.

In order to convert radian measure into degrees, multiply by $\dfrac{180}{\pi}$.

12. $2\pi^R \cdot \dfrac{180}{\pi} = \dfrac{180\pi}{\pi} = 180^o$

14. $\dfrac{2\pi}{3}^R \cdot \dfrac{180}{\pi} = \dfrac{360\pi}{3\pi} = 120^o$

18. $-4\pi^R \cdot \dfrac{180}{\pi} = \dfrac{-720\pi}{\pi} = -720^o$

Given the radius and central angle of each sector, find the arc length and area. All radii are measured in inches.

The formula for arc length is $s = \theta r$ where r is the radius and θ is the measure of the central arc. Keep in mind that θ must be measured in radians. The formula for area is $A = \dfrac{1}{2}\theta r^2$.

22.

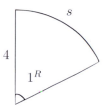

We have $r = 4$ and $\theta = 1^R$

$s = \theta r = (1)(4) = 4$ inches

$A = \dfrac{1}{2}\theta r^2 = \dfrac{1}{2}(1)(4^2) = 8$ square inches

24.

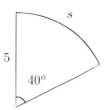

We have $r = 5$ and $\theta = 40^o = \dfrac{2\pi}{9}^R$

$s = \theta r = \left(\dfrac{2\pi}{9}\right)(5) = \dfrac{10\pi}{9}$ inches

$A = \dfrac{1}{2}\theta r^2 = \dfrac{1}{2}\left(\dfrac{2\pi}{9}\right)(5^2) = \dfrac{25\pi}{9}$ square inches

Given the radius and arc length in each diagram, find the measure of the central angle (in radians).

To find the central angle given radius and arc length, we modify the arc length formula as $\theta = \frac{s}{r}$.

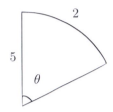

28.

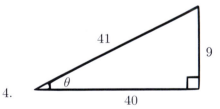

30.

We have $r = 2$ and $s = 5$:

$$\theta = \frac{s}{r} = \frac{5}{2} \text{ radians}$$

We have $r = 4$ and $s = 6$:

$$s = \theta r = \frac{s}{r} = \frac{3}{2} \text{ radians}$$

5.2 Exercises

For exercises 1 - 6, calculate $\sin\theta$, $\cos\theta$ and $\tan\theta$.

Using SOHCAHTOA, we have

$$\sin\theta = \frac{\text{opposite}}{\text{hypotenuse}}, \quad \cos\theta = \frac{\text{adjacent}}{\text{hypotenuse}}, \quad \tan\theta = \frac{\text{opposite}}{\text{adjacent}}$$

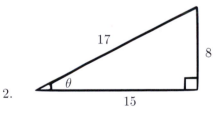

2.

4.

$$\sin\theta = \frac{8}{17}$$

$$\cos\theta = \frac{15}{17}$$

$$\tan\theta = \frac{8}{15}$$

$$\sin\theta = \frac{9}{41}$$

$$\cos\theta = \frac{40}{41}$$

$$\tan\theta = \frac{9}{40}$$

Using the identity $\sin^2\theta + \cos^2\theta = 1$ *and the information given, calculate* $\sin\theta$, $\cos\theta$ *and* $\tan\theta$ *for each problem.*

8. We are given $\sin\theta = 0.8$. Since $\sin^2\theta + \cos^2\theta = 1$, we have $(0.8)^2 + \cos^2\theta = 1$. Therefore $\cos^2\theta = 1 - 0.64 = 0.36$ and $\cos\theta = 0.6$. Since $\tan\theta = \dfrac{\sin\theta}{\cos\theta}$, we have $\tan\theta = \dfrac{0.8}{0.6} = \dfrac{4}{3}$.

10. We are given $\cos\theta = \dfrac{15}{17}$. Since $\sin^2\theta + \cos^2\theta = 1$, we have $\sin^2\theta + \left(\dfrac{15}{17}\right)^2 = 1$. Therefore $\sin^2\theta = 1 - \dfrac{225}{289} = \dfrac{64}{289}$ and $\sin\theta = \dfrac{8}{17}$. Since $\tan\theta = \dfrac{\sin\theta}{\cos\theta}$, we have $\tan\theta = \dfrac{\frac{8}{17}}{\frac{15}{17}} = \dfrac{8}{15}$.

For problems 15 - 22, create a right triangle that satisfies the condition given. Then calculate $\sin\theta$, $\cos\theta$ *and* $\tan\theta$ *for each problem.*

16. $\cos\theta = 0.96 = \dfrac{24}{25} = \dfrac{\text{adjacent}}{\text{hypotenuse}}$. A triangle appears below:

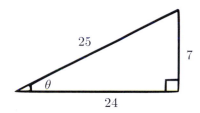

We obtained the third side by using the Pythagorean theorem. We have

$$\sin\theta = \frac{7}{25}, \ \cos\theta = \frac{24}{25}, \ \tan\theta = \frac{7}{24}$$

18. $\tan\theta = \dfrac{5}{12} = \dfrac{\text{opposite}}{\text{adjacent}}$. A triangle appears below:

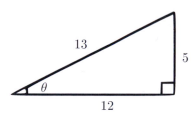

We obtained the hypotenuse by using the Pythagorean theorem. We have

$$\sin\theta = \frac{5}{13}, \ \cos\theta = \frac{12}{13}, \ \tan\theta = \frac{5}{12}$$

In the following, find a value of θ that satisfies the given equation without using a calculator. Answers should be in radians.

24. $\cos\theta = \dfrac{\sqrt{3}}{2}$

    We know that $\cos 30^o = \dfrac{\sqrt{3}}{2}$. 30° is the same as $\dfrac{\pi^R}{6}$.

26. $\sin\theta = 0.5$

    We know that $\sin 30^o = 0.5$. 30° is the same as $\dfrac{\pi^R}{6}$.

28. $\tan\theta = \sqrt{3}$

    We know that $\tan 60^o = \sqrt{3}$. 60° is the same as $\dfrac{\pi^R}{3}$.

5.3   Exercises

For problems 1 - 10, calculate $\cot\theta$, $\sec\theta$ and $\csc\theta$.

2.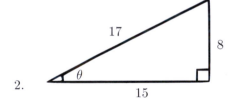

$$\cot\theta = \frac{15}{8}$$

$$\sec\theta = \frac{17}{15}$$

$$\csc\theta = \frac{17}{8}$$

4.

$$\cot\theta = \frac{40}{9}$$

$$\sec\theta = \frac{41}{40}$$

$$\csc\theta = \frac{41}{9}$$

Given the value of the trigonometric function, calculate $\tan\theta$ and $\sec\theta$ using the identity $\tan^2\theta + 1 = \sec^2\theta$.

12. We are given $\sec\theta = 1.25$ so $\tan^2\theta + 1 = 1.25^2$. This means that $\tan^2\theta + 1 = 1.5625$ and $\tan^2\theta = 0.5625$, so $\tan\theta = \sqrt{0.5625} = 1.25$

14.   We have $\tan\theta = \dfrac{5}{12}$, so $\left(\dfrac{5}{12}\right)^2 + 1 = \sec^2\theta$. This leads to $\dfrac{25}{144} + 1 = \sec^2\theta$, so $\sec^2\theta = \dfrac{149}{144}$ and $\sec\theta = \dfrac{13}{12}$.

Given the value of the trigonometric function, calculate $\cot\theta$ *and* $\csc\theta$ *using the identity* $\cot^2\theta + 1 = \csc^2\theta$.

18. $\csc\theta = 1.25$ 20. $\cot\theta = 0.75$

We have $\cot^2\theta + 1 = (1.25)^2$ We have $(0.75)^2 + 1 = \csc^2\theta$
$\cot^2\theta + 1 = 1.5625$ $.5625 + 1 = \csc^2\theta$
$\cot^2\theta = 0.5625$ $\csc^2\theta = 1.5625$
$\cot\theta = 0.75$ $\csc\theta = 1.25$

In each of the following problems, construct a right triangle that satisfies the given condition. Then calculate $\sec\theta$, $\csc\theta$ *and* $\cot\theta$.

24. We are given $\cot\theta = \dfrac{4}{3}$. Since $\cot\theta = \dfrac{\text{adjacent}}{\text{opposite}}$, a triangle representing this case is

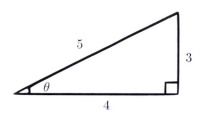

We obtained the hypotenuse by using the Pythagorean theorem. We have

$$\sin\theta = \frac{7}{25}, \ \cos\theta = \frac{24}{25}, \ \tan\theta = \frac{7}{24}$$

Therefore

$$\sec\theta = \frac{25}{24}, \ \csc\theta = \frac{25}{7}, \ \cot\theta = \frac{24}{7}$$

26. We are given $\sec\theta = \dfrac{13}{12}$. Since $\sec\theta = \dfrac{\text{hypotenuse}}{\text{adjacent}}$, a triangle representing this case appears below:

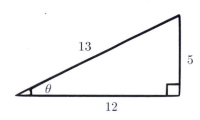

We obtained the third side by using the Pythagorean theorem. We have

$$\sin\theta = \frac{5}{13}, \ \cos\theta = \frac{12}{13}, \ \tan\theta = \frac{5}{12}$$

Therefore

$$\sec\theta = \frac{13}{12}, \ \csc\theta = \frac{13}{5}, \ \cot\theta = \frac{12}{5}$$

In the following, find a value of θ that satisfies the given equation without using a calculator. Answers should be in radians.

30. We are given $\sec\theta = \frac{2\sqrt{3}}{3}$. Therefore $\cos\theta = \frac{3}{2\sqrt{3}} = \frac{\sqrt{3}}{2}$ and $\theta = 60^o$, which is $\frac{\pi}{3}$ radians.

32. We are given $\csc\theta = \sqrt{2}$. Therefore $\sin\theta = \frac{1}{\sqrt{2}} = \frac{\sqrt{2}}{2}$ and $\theta = 45^o$, which is $\frac{\pi}{4}$ radians.

34. We are given $\cot\theta = 1$, so $\tan\theta = 1$ and $\theta = 45^o$, which is $\frac{\pi}{4}$ radians.

5.4   Exercises

Find the reference angle and quadrant of each of the following angles.

Given an angle θ, the angle is in the first quadrant if $0 \le \theta \le 90$, second quadrant if $90 \le \theta \le 180$, third quadrant if $180 \le \theta \le 270$, and fourth quadrant if $270 \le \theta \le 360$. For quadrant II the reference angle is $180 - \theta$. For quadrant III the reference angle is $\theta - 180$. For quadrant IV the reference angle is $360 - \theta$.

2. If $\theta = 300^o$ then θ is in the fourth quadrant and the reference angle is $360 - 300 = 60^o$.

6. If $\theta = 690^o$ then we subtract 360° until it is less than 360°. So a coterminal angle is $690^o - 360^o = 330^o$, which is in the fourth quadrant. The reference angle is $360 - 330 = 30^o$.

8. If $\theta = 420^o$ then the coterminal angle is $420 - 360 = 60^o$, which is in the first quadrant. It is its own reference angle.

12. $\frac{5\pi}{6}^R = 150^o$, which is in the second quadrant. The reference angle for this angle is $180 - 150 = 30^o = \frac{\pi}{6}^R$.

18. $-\frac{7\pi}{6}^R = -210^0$. A coterminal angle is $-210 + 360 = 150^o$, which is in the second quadrant. The reference angle is $180 - 150 = 30^o = \frac{\pi}{6}^R$.

Calculate each of the following without using a calculator:

The steps for finding trigonometric functions of angles outside the first quadrant are:

1. Determine the quadrant the angle lies in.

2. Find the reference angle.

3. Calculate the trigonometric function of the reference angle.

4. Determine the sign of the answer.

22. $\sin 300^o$

This angle lies in the fourth quadrant, and its reference angle is 60°. We have $\sin 60 = \dfrac{\sqrt{3}}{2}$ and sine is negative in the fourth quadrant. Therefore $\sin 300^o = -\dfrac{\sqrt{3}}{2}$.

24. $\cos 345^o$

This angle lies in the fourth quadrant, and its reference angle is 45°. We have $\cos 45 = \dfrac{\sqrt{2}}{2}$ and cosine is positive in the fourth quadrant. Therefore $\cos 345^o = \dfrac{\sqrt{2}}{2}$.

32. $\tan \dfrac{5\pi}{6}^R$

We have $\dfrac{5\pi}{6} = 150^o$. This angle is in the second quadrant and its reference angle is 30°. We have $\tan 30^o = \dfrac{\sqrt{3}}{3}$ and tangent is negative in the second quadrant. Therefore $\tan \dfrac{5\pi}{6}^R = -\dfrac{\sqrt{3}}{3}$.

38. $\tan \left(-\dfrac{7\pi}{6}^R \right)$

We have $-\dfrac{7\pi}{6} = -210^o$. This is in the second quadrant and its reference angle is 30°. Tangent is negative in the second quadrant, so $\tan \left(-\dfrac{7\pi}{6}^R \right) = -\tan 30^o = -\dfrac{\sqrt{3}}{3}$.

Using the information given, determine the quadrant that θ lies in.

Remember that All Students Take Calculus, which indicates which functions are positive or negative in each quadrant. All three are positive in the first quadrant, sine is positive in the second quadrant, tangent is positive in the third quadrant, and cosine is positive in the fourth quadrant.

42. $\tan\theta < 0$ and $\sin\theta < 0$

This means that cosine is positive, so θ is in the fourth quadrant.

44. $\tan\theta > 0$ and $\cos\theta > 0$

Since two of the functions are positive, the third one (sine) must be positive also. Therefore all three are positive, and θ lies in the first quadrant.

46. $\tan\theta < 0$ and $\cos\theta < 0$

In this case two functions are negative so the third one (sine) must be positive. This is true in the second quadrant.

5.5   Exercises

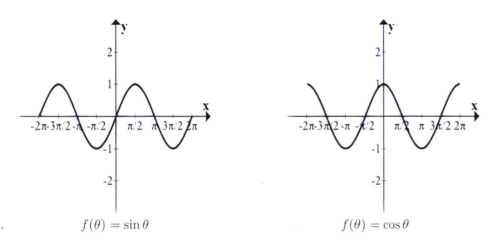

$$f(\theta) = \sin\theta \qquad\qquad f(\theta) = \cos\theta$$

Refer to graphs of the corresponding trigonometric function to answer each of the questions in problems 1 - 10.

2. What is the y-intercept of $f(\theta) = \cos\theta$? The y-intercept is $(0,1)$.

4. What are the x-intercepts of $g(\theta) = \cos\theta$ between -2π and 2π radians? The x-intercepts are $\left(-\dfrac{3\pi}{2},0\right), \left(-\dfrac{\pi}{2},0\right), \left(\dfrac{\pi}{2},0\right), \left(\dfrac{3\pi}{2},0\right)$.

6. For what numbers x, $-2\pi \le x \le 2\pi$, does $\cos x = 1$? $x = -2\pi, 0, 2\pi$

8. What is the range of the function $f(x) = \cos x$? The range of this function is $[-1,1]$.

10. For what numbers x, $-2\pi \le x \le 2\pi$, is the graph of $y = \cos x$ increasing? The function increases on the intervals $(-\pi,0)$ and $(\pi,2\pi)$.

Find the period and amplitude of the following functions.

The period of sine and cosine is 2π radians, and the amplitude is 1. However if we modify the function: $f(\theta) = a\sin b\theta$ or $g(\theta) = a\cos b\theta$, then the period becomes $\dfrac{2\pi}{b}$ radians and the amplitude is a.

12. $g(\theta) = 4\cos(3\theta)$

The period is $\dfrac{2\pi}{3}$ radians and the amplitude is 4.

14. $h(\theta) = 3 - \sin(3\theta)$

The period is $\dfrac{2\pi}{3}$ radians and the amplitude is 1 (amplitude is always positive).

16. $f(\theta) = 3 + 2\sin(2\pi\theta + 4)$

The period is $\dfrac{2\pi}{2} = \pi$ radians and the amplitude is 2.

18. $h(\theta) = 2 + 3\cos(2\theta - \pi)$

The period is $\dfrac{2\pi}{2} = \pi$ radians and the amplitude is 3.

Graph two periods of each of the following functions.

Answers may vary!

20. $g(\theta) = 4\cos(3\theta)$

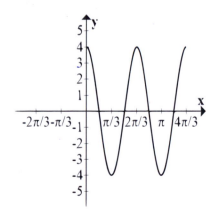

22. $h(\theta) = 3 - \sin(3\theta)$

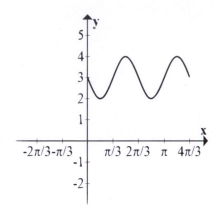

28. $h(\theta) = 2\cos(\theta - 45)$

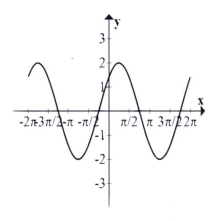

5.6   Exercises

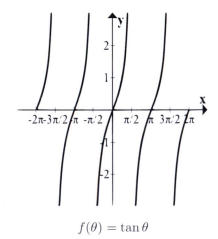
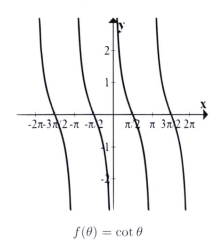

$f(\theta) = \tan\theta$ $f(\theta) = \cot\theta$

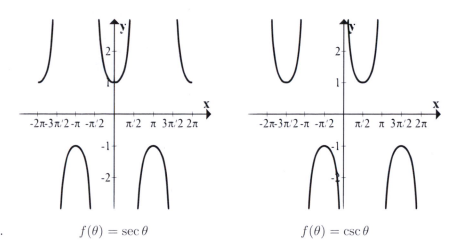

$$f(\theta) = \sec\theta \qquad\qquad\qquad f(\theta) = \csc\theta$$

Refer to graphs of the corresponding trigonometric function to answer each of the questions in problems 1 - 10.

2. What is the y-intercept of $f(\theta) = \cot\theta$? This graph does not have a y-intercept.

4. What is the y-intercept of $g(\theta) = \csc\theta$? This graph does not have a y-intercept.

6. For what numbers x, $-2\pi \le x \le 2\pi$, does $\csc x = 1$? The x values $-\dfrac{3\pi}{2}$ and $\dfrac{\pi}{2}$

8. For what numbers x, $-2\pi \le x \le 2\pi$, does the graph of $y = \cot x$ have vertical asymptotes? $x = -2\pi,\ -\pi,\ 0,\ \pi,\ 2\pi$

10. For what numbers x, $-2\pi \le x \le 2\pi$, does the graph of $y = \csc x$ have vertical asymptotes? $x = -2\pi,\ -\pi,\ 0,\ \pi,\ 2\pi$

In exercises 11 - 20, sketch two full periods of the graph of each function.

Answers may vary!

12. $g(x) = \sec\left(x + \dfrac{\pi}{3}\right)$

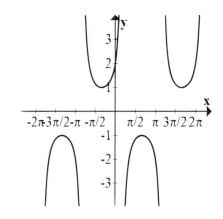

22. $h(x) = \dfrac{1}{2} \csc\left(x - \dfrac{\pi}{2}\right)$

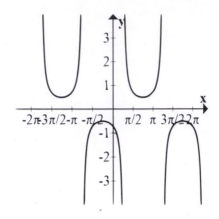

Chapter 6

Trigonometric Identities

6.1 Exercises

Calculate each of the following without using a calculator.

For these problems we will use the following identities:

$$\sin(\alpha + \beta) = \sin\alpha\cos\beta + \cos\alpha\sin\beta$$
$$\sin(\alpha - \beta) = \sin\alpha\cos\beta - \cos\alpha\sin\beta$$
$$\cos(\alpha + \beta) = \cos\alpha\cos\beta - \sin\alpha\sin\beta$$
$$\cos(\alpha - \beta) = \cos\alpha\cos\beta + \sin\alpha\sin\beta$$
$$\tan(\alpha + \beta) = \frac{\tan\alpha + \tan\beta}{1 - \tan\alpha\tan\beta}$$
$$\tan(\alpha - \beta) = \frac{\tan\alpha - \tan\beta}{1 + \tan\alpha\tan\beta}$$

2.

$$
\begin{aligned}
\cos(75^o) &= \cos(45^o + 30^o) \\
&= \cos 45 \cos 30 - \sin 45 \sin 30 \\
&= \left(\frac{\sqrt{2}}{2}\right)\left(\frac{\sqrt{3}}{2}\right) - \left(\frac{\sqrt{2}}{2}\right)\left(\frac{1}{2}\right) \\
&= \frac{\sqrt{6} - \sqrt{2}}{4}
\end{aligned}
$$

3.

$$
\begin{aligned}
\cos(15^o) &= \cos(45^o - 30^o) \\
&= \cos 45 \cos 30 + \sin 45 \sin 30 \\
&= \left(\frac{\sqrt{2}}{2}\right)\left(\frac{\sqrt{3}}{2}\right) + \left(\frac{\sqrt{2}}{2}\right)\left(\frac{1}{2}\right) \\
&= \frac{\sqrt{6} + \sqrt{2}}{4}
\end{aligned}
$$

6.

$$\begin{aligned}
\tan(15^o) &= \tan(45^o - 30^o) \\
&= \frac{\tan 45 - \tan 30}{1 + \tan 45 \tan 30} \\
&= \frac{1 - \frac{\sqrt{3}}{3}}{1 + (1)\left(\frac{\sqrt{3}}{3}\right)} \cdot \frac{3}{3} \\
&= \frac{3 - \sqrt{3}}{3 + \sqrt{3}} \cdot \frac{3 - \sqrt{3}}{3 - \sqrt{3}} \\
&= \frac{9 - 6\sqrt{3} + 3}{9 - 3} \\
&= \frac{12 - 6\sqrt{3}}{6} \\
&= 2 - \sqrt{3}
\end{aligned}$$

12.

$$\begin{aligned}
\cos(105^o) &= \cos(60^o + 45^o) \\
&= \cos 60 \cos 45 - \sin 60 \sin 45 \\
&= \left(\frac{1}{2}\right)\left(\frac{\sqrt{2}}{2}\right) - \left(\frac{\sqrt{3}}{2}\right)\left(\frac{\sqrt{2}}{2}\right) \\
&= \frac{\sqrt{2} - \sqrt{6}}{4}
\end{aligned}$$

In exercises 21 - 26, find the exact value of each function given $\sin s = \frac{3}{5}$ *and* $\cos t = \frac{5}{13}$. *Both angles are in the first quadrant.*

We will use the same identities from above, along with $\sin^2 \alpha + \cos^2 \alpha = 1$. Since $\sin s = \frac{3}{5}$ we can calculate $\cos s = \frac{4}{5}$. Since $\cos t = \frac{5}{13}$ we can calcualte $\sin t = \frac{12}{13}$. Furthermore, $\tan s = \frac{\sin s}{\cos s} = \frac{3}{4}$ and $\tan t = \frac{\sin t}{\cos t} = \frac{12}{5}$.

22.

$$\begin{aligned}
\sin(s - t) &= \sin s \cos t - \cos s \sin t \\
&= \left(\frac{3}{5}\right)\left(\frac{5}{13}\right) - \left(\frac{4}{5}\right)\left(\frac{12}{13}\right) \\
&= \frac{15}{39} - \frac{48}{39} \\
&= -\frac{11}{13}
\end{aligned}$$

24.

$$\begin{aligned}
\cos(s+t) &= \cos s \cos t - \sin s \sin t \\
&= \left(\frac{4}{5}\right)\left(\frac{5}{13}\right) - \left(\frac{3}{5}\right)\left(\frac{12}{13}\right) \\
&= \frac{20}{39} - \frac{36}{39} \\
&= -\frac{16}{13}
\end{aligned}$$

26.

$$\begin{aligned}
\tan(s-t) &= \frac{\tan s - \tan t}{1 + \tan s \tan t} \\
&= \frac{\frac{3}{4} - \frac{12}{5}}{1 + \left(\frac{3}{4}\right)\left(\frac{12}{5}\right)} \cdot \frac{20}{20} \\
&= \frac{15 - 48}{20 + 36} \\
&= -\frac{33}{56}
\end{aligned}$$

Verify each of the following identities.

When verifying an identity, we start with one side of the equation. Then apply previously known formulas, one at a time, and manipulate the equation until it matches the other side.

28.

$$\begin{aligned}
\cos(x+y) + \cos(x-y) &= 2\cos x \cos y \\
(\cos x \cos y - \sin x \sin y) + (\cos x \cos y + \sin x \sin y) &= 2\cos x \cos y \\
\cos x \cos y - \sin x \sin y + \cos x \cos y + \sin x \sin y &= 2\cos x \cos y \\
2\cos x \cos y &= 2\cos x \cos y
\end{aligned}$$

30.

$$\begin{aligned}
\sin(x+y)\sin(x-y) &= \sin^2 x - \sin^2 y \\
(\sin x \cos y + \cos x \sin y)(\sin x \cos y - \cos x \sin y) &= \sin^2 x - \sin^2 y \\
\sin^2 x \cos^2 y - \sin x \cos x \sin y \cos y + \sin x \cos x \sin y \cos y - \cos^2 x \sin^2 y &= \sin^2 x - \sin^2 y \\
\sin^2 x \cos^2 y - \cos^2 x \sin^2 y &= \sin^2 x - \sin^2 y \\
\sin^2 x \cos^2 y + \sin^2 x \sin^2 y - \sin^2 x \sin^2 y - \cos^2 x \sin^2 y &= \sin^2 x - \sin^2 y \\
(\sin^2 x \cos^2 y + \sin^2 x \sin^2 y) - (\sin^2 x \sin^2 y + \cos^2 x \sin^2 y) &= \sin^2 x - \sin^2 y \\
\sin^2 x (\cos^2 y + \sin^2 y) - \sin^2 y (\sin^2 x + \cos^2 x) &= \sin^2 x - \sin^2 y \\
\sin^2 x (1) - \sin^2 y (1) &= \sin^2 x - \sin^2 y
\end{aligned}$$

32. $\cot(\alpha - \beta) = \dfrac{\cot\alpha\cot\beta + 1}{\cot\beta - \cot\alpha}$

$$
\begin{aligned}
\cot(\alpha - \beta) &= \frac{\cos(\alpha - \beta)}{\sin(\alpha - \beta)} \\[2mm]
&= \frac{\cos\alpha\cos\beta + \sin\alpha\sin\beta}{\sin\alpha\cos\beta - \cos\alpha\sin\beta} \cdot \frac{\frac{1}{\sin\alpha\sin\beta}}{\frac{1}{\sin\alpha\sin\beta}} \\[2mm]
&= \frac{\frac{\cos\alpha\cos\beta}{\sin\alpha\sin\beta} + \frac{\sin\alpha\sin\beta}{\sin\alpha\sin\beta}}{\frac{\sin\alpha\cos\beta}{\sin\alpha\sin\beta} - \frac{\cos\alpha\sin\beta}{\sin\alpha\sin\beta}} \\[2mm]
&= \frac{\left(\frac{\cos\alpha}{\sin\alpha}\right)\left(\frac{\cos\beta}{\sin\beta}\right) + 1}{\frac{\cos\beta}{\sin\beta} - \frac{\cos\alpha}{\sin\alpha}} \\[2mm]
&= \frac{\cot\alpha + 1}{\cot\beta - \cot\alpha}
\end{aligned}
$$

34. $\sec(\alpha - \beta) = \dfrac{\sec\alpha\sec\beta}{1 + \tan\alpha\tan\beta}$

$$
\begin{aligned}
\sec(\alpha - \beta) &= \frac{1}{\cos(\alpha - \beta)} \\[2mm]
&= \frac{1}{\cos\alpha\cos\beta + \sin\alpha\sin\beta} \cdot \frac{\frac{1}{\cos\alpha\cos\beta}}{\frac{1}{\cos\alpha\cos\beta}} \\[2mm]
&= \frac{\left(\frac{1}{\cos\alpha}\right)\left(\frac{1}{\cos\beta}\right)}{\frac{\cos\alpha\cos\beta}{\cos\alpha\cos\beta} - \frac{\sin\alpha\sin\beta}{\cos\alpha\cos\beta}} \\[2mm]
&= \frac{\left(\frac{1}{\cos\alpha}\right)\left(\frac{1}{\cos\beta}\right)}{1 - \left(\frac{\sin\alpha}{\cos\alpha}\right)\left(\frac{\sin\beta}{\cos\beta}\right)} \\[2mm]
&= \frac{\sec\alpha\sec\beta}{1 + \tan\alpha\tan\beta}
\end{aligned}
$$

Find all solutions between 0 and 360 degrees to each of the following equations.

Here we use identities to simplify the right hand side.

38.

$$\begin{aligned}
\cos(x+45) - \cos(x-45) &= 1 \\
(\cos x \cos 45 - \sin x \sin 45) - (\cos x \cos 45 + \sin x \sin 45) &= 1 \\
\cos x \cos 45 - \sin x \sin 45 - \cos x \cos 45 - \sin x \sin 45 &= 1 \\
-2 \sin x \sin 45 &= 1 \\
(-2 \sin x)\left(\frac{\sqrt{2}}{2}\right) &= 1 \\
-\sqrt{2} \sin x &= 1 \\
\sin x &= -\frac{1}{\sqrt{2}} \\
\sin x &= -\frac{\sqrt{2}}{2} \\
x &= 225^\circ,\ 315^\circ
\end{aligned}$$

40.

$$\begin{aligned}
2\sin(x+90) + 3\tan(180-x) &= 0 \\
2(\sin x \cos 90 + \cos x \sin 90) + 3\left(\frac{\tan 180 - \tan x}{1 + \tan 180 \tan x}\right) &= 0 \\
2\cos x - \frac{3\tan x}{1} &= 0 \\
2\cos x - 3\tan x &= 0 \\
2\cos x - \frac{3\sin x}{\cos x} &= 0 \\
\cos x \left(2\cos x - \frac{3\sin x}{\cos x}\right) &= 0 \\
2\cos^2 x - 3\sin x &= 0 \\
2(1 - \sin^2 x) - 3\sin x &= 0 \\
2 - 2\sin^2 x - 3\sin x &= 0 \\
2\sin^2 x + 3\sin x - 2 &= 0 \\
(2\sin x - 1)(\sin x + 2) &= 0
\end{aligned}$$

This gives the equations $2\sin x - 1 = 0$ and $\sin x + 2 = 0$. The first equation leads to $\sin x = \frac{1}{2}$. The solutions to this equation are $x = 30^\circ,\ 150^\circ$. The second equation gives $\sin x = -2$, which has no solutions.

6.2   Exercises

Calculate the exact value of each of the following using an appropriate half angle formula.

The half angle formulas for sine, cosine, and tangent are

$$\sin\frac{\theta}{2} = \pm\sqrt{\frac{1-\cos\theta}{2}} \qquad \cos\frac{\theta}{2} = \pm\sqrt{\frac{1+\cos\theta}{2}}$$

$$\tan\frac{\theta}{2} = \sqrt{\frac{1-\cos\theta}{1+\cos\theta}} = \frac{\sin\theta}{1+\cos\theta} = \frac{1-\cos\theta}{\sin\theta}$$

2. $\cos 15^\circ = \sqrt{\dfrac{1+\cos 30}{2}} = \sqrt{\dfrac{1+\frac{\sqrt{3}}{2}}{2}} = \sqrt{\dfrac{2+\sqrt{3}}{4}} = \dfrac{\sqrt{2+\sqrt{3}}}{2}$

4. $\sin 75^\circ = \sqrt{\dfrac{1-\cos 150}{2}} = \sqrt{\dfrac{1-\left(-\frac{\sqrt{3}}{2}\right)}{2}} = \sqrt{\dfrac{2+\sqrt{3}}{4}} = \dfrac{\sqrt{2+\sqrt{3}}}{2}$

6. $\tan 75^\circ = \sqrt{\dfrac{1-\cos 150}{1+\cos 150}} = \sqrt{\dfrac{1-\left(-\frac{\sqrt{3}}{2}\right)}{1+\left(-\frac{\sqrt{3}}{2}\right)}} = \sqrt{\dfrac{2+\sqrt{3}}{2-\sqrt{3}}\cdot\dfrac{2+\sqrt{3}}{2+\sqrt{3}}} = \sqrt{\dfrac{(2+\sqrt{3})^2}{4-3}} = 2+\sqrt{3}$

Verify each of the following identities.

When verifying an identity, we start with one side of the equation. Then apply previously known formulas, one at a time, and manipulate the equation until it matches the other side.

12.

$$
\begin{aligned}
\cos 3\theta &= \cos(2\theta+\theta) \\
&= \cos 2\theta\cos\theta - \sin 2\theta\sin\theta \\
&= (2\cos^2\theta - 1)\cos\theta - (2\sin\theta\cos\theta)\sin\theta \\
&= 2\cos^3\theta - \cos\theta - 2\sin^2\theta\cos\theta \\
&= 2\cos^3\theta - \cos\theta - 2(1-\cos^2\theta)\cos\theta \\
&= 2\cos^3\theta - \cos\theta - 2\cos\theta + 2\cos^3\theta \\
&= 4\cos^3\theta - 3\cos\theta
\end{aligned}
$$

14.

$$
\begin{aligned}
\cos 4\theta &= \cos(2 \cdot 2\theta) \\
&= 2\cos^2 2\theta - 1 \\
&= 2(2\cos^2 \theta - 1)^2 - 1 \\
&= 2(2\cos^2 \theta - 1)(2\cos^2 \theta - 1) - 1 \\
&= 2(4\cos^4 \theta - 2\cos^2 \theta - 2\cos^2 \theta + 1) - 1 \\
&= 8\cos^4 \theta - 8\cos^2 \theta + 1
\end{aligned}
$$

Using the diagram below, calculate the value of each trigonometric function.

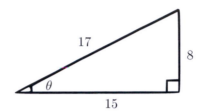

16. $\cos \dfrac{\theta}{2} = \sqrt{\dfrac{1+\cos\theta}{2}} = \sqrt{\dfrac{1+\frac{15}{17}}{2}} = \sqrt{\dfrac{17+15}{2(17)}} = \sqrt{\dfrac{16}{17}} = \dfrac{4\sqrt{17}}{17}$

18. $\cot \dfrac{\theta}{2} = \dfrac{\sin\theta}{1-\cos\theta} = \dfrac{\frac{8}{17}}{1-\frac{15}{17}} \cdot \dfrac{17}{17} = \dfrac{8}{17-15} = 4$

24. $\tan \dfrac{\theta}{2} + \cot \dfrac{\theta}{2} = \dfrac{1}{4} + 4 = \dfrac{17}{4}$ (from problem 18)

Given the fact that $\cos\alpha = \dfrac{7}{25}$ and $\sin\beta = \dfrac{3}{5}$ and that α and β are acute angles, calculate each of the following:

We use the fact that $\sin\beta = \dfrac{3}{5}$ and the identity $\cos^2\beta + \sin^2\beta = 1$ to obtain $\cos\beta = \dfrac{4}{5}$.

26. $\sin \dfrac{\alpha}{2} = \sqrt{\dfrac{1-\cos\alpha}{2}} = \sqrt{\dfrac{1-\frac{7}{25}}{2}} = \sqrt{\dfrac{25-7}{2(25)}} = \sqrt{\dfrac{9}{25}} = \dfrac{3}{5}$

28. $\tan \dfrac{\beta}{2} = \dfrac{\sin\beta}{1+\cos\beta} = \dfrac{\frac{3}{5}}{1+\frac{4}{5}} \cdot \dfrac{5}{5} = \dfrac{3}{5+4} = \dfrac{1}{3}$

30. $\cos \dfrac{\beta}{2} = \sqrt{\dfrac{1+\cos\beta}{2}} = \sqrt{\dfrac{1+\frac{4}{5}}{2}} \cdot \dfrac{5}{5} = \sqrt{\dfrac{5+4}{10}} = \sqrt{\dfrac{9}{10}} = \dfrac{3\sqrt{10}}{10}$

6.3   Exercises

In problems 1 - 10, calculate each value. Write the answers in degrees.

Remember that arcsin is another way to write inverse sine. So if $y = \arcsin x$, then $x = \sin y$.

Similarly, if $y = \arccos x$ then $x = \cos y$, and if $y = \arctan x$ then $x = \tan y$.

2. Let $y = \arccos\left(\dfrac{\sqrt{3}}{2}\right)$. Then $\cos y = \dfrac{\sqrt{3}}{2}$, and $y = 30^o$.

4. Let $y = \arctan(-1)$. Then $\tan y = -1$, and $y = -45^o$. Remember that the range of the arctangent function is $(-90^o, 90^o)$.

6. In order to evaluate $\sin(\arccos .5)$, we will let $y = \arccos .5$. Then $\cos y = .5$ and $y = 60^o$. Therefore $\sin(\arccos .5) = \sin 60^o = \dfrac{\sqrt{3}}{2}$.

8. In order to evaluate $\arccos(\sin 30)$, we first calculate $\sin 30 = \dfrac{1}{2}$. Then let $y = \arccos(\sin 30) = \arccos\left(\dfrac{1}{2}\right)$, and $\cos y = \dfrac{1}{2}$, so $y = 60^o$.

10. In order to evaluate $\cos(\arctan 1)$, we let $y = \arctan 1$. Then $\tan y = 1$ and $y = 45^o$. Therefore $\cos(\arctan 1) = \cos y = \cos 45^o = \dfrac{\sqrt{2}}{2}$.

In problems 11 - 20, find each value exactly. Write the answers in degrees.

14. Let $y = \arccos(0.5)$. Then $\cos y = 0.5$ and $y = 60^o$.

16. Let $y = \csc^{-1}\left(\dfrac{2\sqrt{3}}{3}\right)$. Then $\csc y = \dfrac{2\sqrt{3}}{3}$ and $\sin y = \dfrac{3}{2\sqrt{3}} = \dfrac{\sqrt{3}}{2}$ and $y = 60^o$.

18. Let $y = \cos^{-1}(-1)$. Then $\cos y = -1$ and $y = -90^o$. Remember that the range of $f(x) = \arccos x$ is $[-90^0, 90^o]$.

20. Let $y = \arccos\left(-\dfrac{1}{2}\right)$. Then $\cos y = -\dfrac{1}{2}$ and $y = -60^o$.

In problems 21 - 30, use a right triangle to simplify each expression.

22. In order to evaluate $\cos(\sin^{-1} x)$, first let $\theta = \sin^{-1} x$. Then $\sin \theta = x$. Since $\sin \theta = \dfrac{\text{opposite}}{\text{hypotenuse}}$, we can let $\sin \theta = \dfrac{x}{1}$, giving the triangle

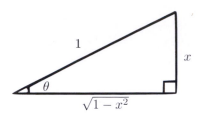

Therefore $\cos(\sin^{-1} x) = \cos \theta = \dfrac{\text{adjacent}}{\text{hypotenuse}} = \sqrt{1 - x^2}$.

24. In order to evaluate $\cot(\csc^{-1} x)$, we define $\theta = \csc^{-1} x$ so that $\csc \theta = x$. Since $\csc \theta = \dfrac{\text{hypotenuse}}{\text{opposite}}$, we have $\csc \theta = \dfrac{x}{1}$, leading to the triangle

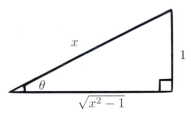

Therefore $\cot(\csc^{-1} x) = \dfrac{\text{adjacent}}{\text{opposite}} = \sqrt{x^2 - 1}$

26. In order to evaluate $\csc\left(\cot^{-1} \dfrac{x}{4}\right)$, we define $\theta = \cot^{-1} \dfrac{x}{4}$, so that $\cot \theta = \dfrac{x}{4} = \dfrac{\text{adjacent}}{\text{opposite}}$. This gives the triangle

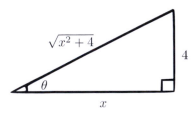

Therefore $\csc\left(\cot^{-1} \dfrac{x}{4}\right) = \dfrac{\text{hypotenuse}}{\text{opposite}} = \dfrac{\sqrt{x^2 + 4}}{4}$.

6.4   Exercises

In exercises 1 - 10, use the Law of Sines to solve each triangle.

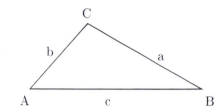

Figure 6.1: Triangle for the Law of Sines

The Law of Sines states that for the above triangle,

$$\frac{\sin A}{a} = \frac{\sin B}{b} = \frac{\sin C}{c}$$

It is useful if we have an angle and opposite side, e.g. angle A and side a.

2. $C = 60^o$, $c = 10$, $B = 78^o$

This is an example of an angle, angle, side triangle. We will use angles C and B along with side c to determine the length of side b:

$$\frac{\sin B}{b} = \frac{\sin C}{c}$$

$$\frac{\sin 78}{b} = \frac{\sin 60}{10}$$

$$b \sin 60 = 10 \sin 78$$

$$b = \frac{10 \sin 78}{\sin 60}$$

$$\approx 11.29$$

Now we use the fact that the sum of the angles in a triangle is 180°, so $A = 180 - 60 - 78 = 42$.

Finally, we need side a:

$$\frac{\sin A}{a} = \frac{\sin C}{c}$$

$$\frac{\sin 42}{a} = \frac{\sin 60}{10}$$

$$a \sin 60 = 10 \sin 42$$

$$a = \frac{10 \sin 42}{\sin 60}$$

$$\approx 7.73$$

4. $A = 75^o$, $a = 8$, $C = 45^o$

This is also an angle, angle, side. We start with angles A and C and side a, then determine side c:

$$\frac{\sin A}{a} = \frac{\sin C}{c}$$

$$\frac{\sin 75}{8} = \frac{\sin 45}{c}$$

$$c \sin 75 = 8 \sin 45$$

$$c = \frac{8 \sin 45}{\sin 75}$$

$$\approx 5.86$$

In exercises 11 - 20, determine the number of solutions for each triangle. Then use the Law of Sines to find each solution.

These problems are all side, side, angle, which is the ambiguous case. We have the following cases:

Different Solutions for SSA

Suppose we are given the angle A and sides a and b. Then the possible number of solutions to the triangle are as follows:

Case I: A is acute

$$a < b \sin A: \text{ no solutions}$$
$$a = b \sin A: \text{ one solution}$$
$$b \sin A < a < b: \text{ two solutions}$$
$$a > b: \text{ one solution}$$

Case II: A is obtuse

$$a \leq b: \text{ no solutions}$$
$$a > b: \text{ one solution}$$

---

12. $A = 70^o$, $a = 12$, $b = 19$

In this case A is acute, $b \sin A \approx 17.854 > 12 = a$. Therefore there is no solution for this triangle.

16. $A = 58^o$, $a = 11$, $b = 7$

Here A is acute and $a > b$, so there is one solution.

$$\frac{\sin A}{a} = \frac{\sin B}{b}$$
$$\frac{\sin 58}{11} = \frac{\sin B}{7}$$
$$\sin B = \frac{7 \sin 58}{11}$$
$$\approx 0.5397$$
$$B \approx 32.66^o$$

The third angle $C = 180 - 70 - 32.66 = 77.34^o$. Using the Law of Sines again:

$$\frac{\sin A}{a} = \frac{\sin C}{c}$$
$$\frac{\sin 58}{11} = \frac{\sin 77.34}{c}$$
$$c \sin 58 = 11 \sin 77.34$$
$$c = \frac{11 \sin 77.34}{\sin 58}$$
$$\approx 12.66$$

18. $A = 39^o$, $a = 10$, $b = 13$

Here $b \sin A \approx 8.18 < 10 = a < 13 = b$, so there are two solutions.

$$\frac{\sin A}{a} = \frac{\sin B}{b}$$
$$\frac{\sin 39}{10} = \frac{\sin B}{13}$$
$$\sin B = \frac{13 \sin 39}{10}$$
$$\approx 0.8181$$

This equation has two solutions that are less than 180°: $B \approx 54.9^o$ and $B = 125.1^o$. Using the first answer, we get a third angle of $C = 180 - 39 - 54.9 = 86.1$. This leads to

$$\frac{\sin A}{a} = \frac{\sin C}{c}$$

$$\frac{\sin 39}{10} = \frac{\sin 86.1}{c}$$

$$c \sin 39 = 10 \sin 86.1$$

$$c = \frac{10 \sin 86.1}{\sin 39}$$

$$\approx 15.85$$

Using the value $B = 125.1$, we get a third angle of $C = 180 - 39 - 125.1 = 15.9$:

$$\frac{\sin A}{a} = \frac{\sin C}{c}$$

$$\frac{\sin 39}{10} = \frac{\sin 15.9}{c}$$

$$c \sin 39 = 10 \sin 15.9$$

$$c = \frac{10 \sin 15.9}{\sin 39}$$

$$\approx 4.35$$

20. $A = 130^o$, $a = 15$, $b = 18$

Here A is obtuse and $a < b$, so there is no solution.

In exercises 21 - 30, calculate the area of each triangle.

For these problems we use the equivalent formulas

$$\text{Area} = \frac{1}{2}ab \sin C = \frac{1}{2}ac \sin B = \frac{1}{2}bc \sin A$$

22. $C = 48^o$, $a = 10$, $b = 15$

$$\text{Area} = \frac{1}{2}ab \sin C = \frac{1}{2}(10)(15) \sin 48 \approx 55.74$$

24. $A = 58^o$, $b = 14$, $c = 8$

$$\text{Area} = \frac{1}{2}bc \sin A = \frac{1}{2}(14)(8) \sin 58 \approx 47.49$$

26. $B = 39^o$, $a = 10$, $c = 13$

$$\text{Area} = \frac{1}{2}ac \sin B = \frac{1}{2}(10)(13) \sin 39 \approx 40.91$$

6.5   Exercises

In problems 1 - 6, solve each triangle.

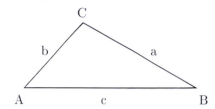

Figure 6.2: Triangle for the Law of Cosines

The law of cosines states that

$$c^2 = a^2 + b^2 - 2ab\cos C$$

We can also vary the order of the sides and angle. The letter denoting the angle is always the same as the letter on the left hand side.

2.

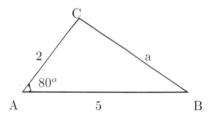

Here we have $b = 2$, $c = 5$, and $A = 80$. This is a side, angle, side case. We will use the equation $a^2 = b^2 + c^2 - 2bc\cos A$:

$$
\begin{aligned}
a^2 &= b^2 + c^2 - 2bc\cos A \\
&= 2^2 + 5^2 - 2(2)(5)\cos 80 \\
&= 29 - 20\cos 80 \\
a &= \sqrt{29 - 20\cos 80} \\
&\approx 5.05
\end{aligned}
$$

Now we will use the Law of Sines to find the angle B, which must be acute:

$$
\begin{aligned}
\frac{\sin A}{a} &= \frac{\sin B}{b} \\
\frac{\sin 80}{5.05} &= \frac{\sin B}{2} \\
\sin B &= \frac{2\sin 80}{5.05} \\
&\approx 0.39 \\
B &\approx 22.96^o
\end{aligned}
$$

To get angle C we calculate $C = 180 - A - B = 180 - 80 - 22.96 = 77.04^o$.

4.

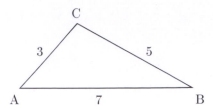

We are given $a = 5$, $b = 3$, and $c = 7$. Using the Law of Cosines $c^2 = a^2 + b^2 - 2ab\cos C$ and solving this equation for $\cos C$, we get

$$\cos C = \frac{a^2 + b^2 - c^2}{2ab}$$

$$\begin{aligned}
\cos C &= \frac{a^2 + b^2 - c^2}{2ab} \\
&= \frac{5^2 + 3^2 - 7^2}{2(5)(3)} \\
&= -\frac{1}{2} \\
C &= 120^o
\end{aligned}$$

Next we will calculate the angle A using the Law of Sines:

$$\begin{aligned}
\frac{\sin A}{a} &= \frac{\sin C}{c} \\
\frac{\sin A}{5} &= \frac{\sin 120}{7} \\
\sin A &= \frac{5\sin 120}{7} \\
&\approx 0.6186 \\
A &\approx 38.2^o
\end{aligned}$$

To calculate the angle B we use $B = 180 - A - C = 180 - 38.2 - 120 = 21.8^o$.

In problems 7 - 16, solve each triangle.

Here we start with the Law of Cosines, then apply the Law of Sines.

8. $a = 1$, $c = 2$, $B = 40^o$

$$\begin{aligned}
b^2 &= a^2 + c^2 - 2ac\cos B \\
&= 1^2 + 2^2 - 2(1)(2)\cos 40 \\
&= 5 - 4\cos 40 \\
b &= \sqrt{5 - 4\cos 40} \\
&\approx 1.39
\end{aligned}$$

Now we will calculate angle A using the Law of Sines:

$$\frac{\sin A}{a} = \frac{\sin B}{b}$$
$$\frac{\sin A}{1} = \frac{\sin 40}{1.39}$$
$$\sin A = \frac{\sin 40}{1.39}$$
$$A \approx 27.5^o$$

To find angle C we use $C = 180 - A - B = 180 - 27.5 - 40 = 112.5^o$.

12. $b = 7$, $c = 3$, $A = 80^o$

$$
\begin{aligned}
a^2 &= b^2 + c^2 - 2bc\cos A \\
&= 7^2 + 3^2 - 2(7)(3)\cos 80 \\
&= 58 - 42\cos 80 \\
a &= \sqrt{58 - 42\cos 80} \\
&\approx 7.12
\end{aligned}
$$

Next we find angle C using the Law of Sines. C must be acute because $c < b$:

$$\frac{\sin A}{a} = \frac{\sin C}{c}$$
$$\frac{\sin 80}{7.12} = \frac{\sin C}{3}$$
$$\sin C = \frac{3\sin 80}{7.12}$$
$$C \approx 24.5^o$$

To get angle C we use $C = 180 - A - B = 180 - 80 - 24.5 = 75.5^o$.

18. The distance from home plate to the fence in dead center field in Fenway Park is exactly 390 feet. How far is it from the fence in dead center to third base (Note: the distance from home plate to third base is 90 feet)?

For this problem, refer to the diagram on the next page. A line from home plate to dead center forms a 45° angle with the line from home plate to third base. If we complete this triangle with a line from dead center to third base, let $a = 90$ (the line from home plate to third base) and $b = 390$ (the line from home plate to dead center). Then $C = 45^o$ (the angle between the dead center line and the third base line) and we wish to know c. We have

$$
\begin{aligned}
c^2 &= a^2 + b^2 - 2ab\cos C \\
&= 90^2 + 390^2 - 2(90)(390)\cos 45 \\
&= 160200 - 70200\cos 45 \\
c &= \sqrt{160200 - 70200\cos 45} \\
&\approx 332.5 \text{ feet}
\end{aligned}
$$

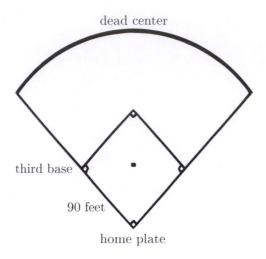

6.6 Exercises

Verify each identity.

When verifying an identity, we start with one side of the equation. Then apply previously known formulas, one at a time, and manipulate the equation until it matches the other side.

2. $\cos\theta\csc\theta = \cot\theta$

We will convert the left hand side to sines and cosines:

$$
\begin{aligned}
\cos\theta\csc\theta &= \cos\theta\left(\frac{1}{\sin\theta}\right) \\
&= \frac{\cos\theta}{\sin\theta} \\
&= \cot\theta
\end{aligned}
$$

4. $\tan\beta\csc\beta\cos\beta = 1$

We will convert the left hand side to sines and cosines:

$$
\begin{aligned}
\tan\beta\csc\beta\cos\beta &= \left(\frac{\sin\beta}{\cos\beta}\right)\left(\frac{1}{\sin\beta}\right)\cos\beta \\
&= \frac{\sin\beta\cos\beta}{\sin\beta\cos\beta} \\
&= 1
\end{aligned}
$$

8. $\tan \alpha = \dfrac{\cos \alpha \sec \alpha}{\cot \alpha}$

We will convert the right hand side to sines and cosines:

$$
\begin{aligned}
\frac{\cos \alpha \sec \alpha}{\cot \alpha} &= \frac{\cos \alpha \left(\frac{1}{\cos \alpha}\right)}{\frac{\cos \alpha}{\sin \alpha}} \\[2mm]
&= \frac{1}{\frac{\cos \alpha}{\sin \alpha}} \\[2mm]
&= \frac{\sin \alpha}{\cos \alpha} \\[2mm]
&= \tan \alpha
\end{aligned}
$$

12. $\dfrac{\cos^2 x - \sin^2 x}{\sin x \cos x} = \cot x - \tan x$

We will break the left hand side into two fractions, then apply two trigonometric identitites:

$$
\begin{aligned}
\frac{\cos^2 x - \sin^2 x}{\sin x \cos x} &= \frac{\cos^2 x}{\sin x \cos x} - \frac{\sin^2 x}{\sin x \cos x} \\[2mm]
&= \frac{\cos x}{\sin x} - \frac{\sin x}{\cos x} \\[2mm]
&= \cot x - \tan x
\end{aligned}
$$

16. $\dfrac{\sin u}{1 - \cos^2 u} = \csc u$

We will apply the identity $\sin^2 u = 1 - \cos^2 u$ on the left hand side, then simplify:

$$
\begin{aligned}
\frac{\sin u}{1 - \cos^2 u} &= \frac{\sin u}{\sin^2 u} \\[2mm]
&= \frac{1}{\sin u} \\[2mm]
&= \csc u
\end{aligned}
$$

18. $(1 - \sin t)(1 + \sin t) = \cos^2 t$

We will start by multiplying out the left hand side:

$$
\begin{aligned}
(1 - \sin t)(1 + \sin t) &= 1 + \sin t - \sin t - \sin^2 t \\
&= 1 - \sin^2 t \\
&= \cos^2 t
\end{aligned}
$$

20. $(\sin x + \cos x)^2 = 1 + \sin 2x$

We will start by multiplying out the left hand side:

$$
\begin{aligned}
(\sin x + \cos x)^2 &= (\sin x + \cos x)(\sin x + \cos x) \\
&= \sin^2 x + \sin x \cos x + \sin x \cos x + \cos^2 x \\
&= (\sin^2 x + \cos^2 x) + 2 \sin x \cos x \\
&= 1 + 2 \sin x \cos x \\
&= 1 + \sin 2x
\end{aligned}
$$

22. $\dfrac{\tan x + \cot y}{\tan x \cot y} = \tan y + \cot x$

We will break the left hand side into two fractions, then simplify:

$$
\begin{aligned}
\frac{\tan x + \cot y}{\tan x \cot y} &= \frac{\tan x}{\tan x \cot y} + \frac{\cot y}{\tan x \cot y} \\
&= \frac{1}{\cot y} + \frac{1}{\tan x} \\
&= \tan y + \cot x
\end{aligned}
$$

6.7 Exercises

In problems 1 - 20, find all solutions between 0 and 360 degrees for each equation.

In these problems we will try to isolate one trigonometric function, then solve for θ.

2.

$$
\begin{aligned}
2\cos\theta + \sqrt{3} &= 0 \\
2\cos\theta &= -\sqrt{3} \\
\cos\theta &= -\frac{\sqrt{3}}{2} \\
\theta &= 150^o,\ 210^o
\end{aligned}
$$

4.

$$
\begin{aligned}
\sqrt{3}\cot\theta - 1 &= 0 \\
\sqrt{3}\cot\theta &= 1 \\
\cot\theta &= \frac{1}{\sqrt{3}} \\
&= \frac{\sqrt{3}}{3} \\
\theta &= 60^o,\ 240^o
\end{aligned}
$$

6.

$$
\begin{aligned}
2\sin 3\theta - 1 &= 0 \\
2\sin 3\theta &= 1 \\
\sin 3\theta &= \frac{1}{2} \\
3\theta &= 30^o,\ 150^o,\ 390^o,\ 510^o,\ 750^o,\ 870^o \\
\theta &= 10^o,\ 50^o,\ 170^o,\ 250^o,\ 290^o
\end{aligned}
$$

8. $(\tan\theta + \sqrt{3})(\sqrt{3}\csc\theta - 2) = 0$

This gets broken into two equations. The first equation is

$$\begin{aligned} \tan\theta + \sqrt{3} &= 0 \\ \tan\theta &= -\sqrt{3} \\ \theta &= 120^o,\ 300^o \end{aligned}$$

The second equation is

$$\begin{aligned} \sqrt{3}\csc\theta - 2 &= 0 \\ \sqrt{3}\csc\theta &= 2 \\ \csc\theta &= \frac{2}{\sqrt{3}} \\ \sin\theta &= \frac{\sqrt{3}}{2} \\ \theta &= 60^o,\ 120^o \end{aligned}$$

Therefore the solutions are $\theta = 60^o,\ 120^o,\ 300^o$.

10.

$$\begin{aligned} \tan\theta &= \sec\theta \\ \frac{\sin\theta}{\cos\theta} &= \frac{1}{\cos\theta} \\ \cos\theta\left(\frac{\sin\theta}{\cos\theta}\right) &= \cos\theta\left(\frac{1}{\cos\theta}\right) \\ \sin\theta &= 1 \\ \theta &= 90^o \end{aligned}$$

However, $\tan\theta$ does not exist when $\theta = 90^0$. Therefore this equation has no solutions.

12.

$$\begin{aligned} \cos 2\theta + \cos\theta &= 0 \\ (2\cos^2\theta - 1) + \cos\theta &= 0 \\ 2\cos^2\theta + \cos\theta - 1 &= 0 \\ (2\cos\theta - 1)(\cos\theta + 1) &= 0 \end{aligned}$$

This breaks into two equations. The first equation:

$$\begin{aligned} 2\cos\theta - 1 &= 0 \\ 2\cos\theta &= 1 \\ \cos\theta &= \frac{1}{2} \\ \theta &= 60^o,\ 300^o \end{aligned}$$

The second equation:

$$\begin{aligned} \cos\theta + 1 &= 0 \\ \cos\theta &= -1 \\ \theta &= 180^o \end{aligned}$$

Therefore the solutions are $\theta = 60°$, $180°$, $300°$.

14.

$$
\begin{aligned}
\sin 4\theta - \sin 8\theta &= 0 \\
\sin 4\theta - \sin 2(4\theta) &= 0 \\
\sin 4\theta - 2\sin 4\theta \cos 4\theta &= 0 \\
\sin 4\theta(1 - 2\cos 4\theta) &= 0
\end{aligned}
$$

This breaks into two equations:

$$
\begin{aligned}
\sin 4\theta &= 0 \\
4\theta &= 0°,\ 180°,\ 360°,\ 540°,\ 720°,\ 800°,\ 1080°,\ 1160°,\ 1440° \\
\theta &= 0°,\ 45°,\ 90°,\ 135°,\ 180°,\ 225°,\ 270°,\ 315°,\ 360°
\end{aligned}
$$

The other equation:

$$
\begin{aligned}
1 - 2\cos 4\theta &= 0 \\
2\cos 4\theta &= 1 \\
\cos 4\theta &= \frac{1}{2} \\
4\theta &= 60°,\ 300°,\ 420°,\ 660°,\ 780°,\ 1020°,\ 1140°,\ 1380° \\
\theta &= 15°,\ 75°,\ 105°,\ 165°,\ 195°,\ 255°,\ 285°,\ 345°
\end{aligned}
$$

In problems 21 - 30, solve each equation using a graphing utility. Give all answers in radians.

To solve these problems using a TI-83, first get a zero on the right hand side. Then create a graph of the function defined by the left hand side, and use the ZERO method to solve.

22. $x - 4\sin x = 0$

A graph of the function $f(x) = x - 4\sin x$ appears below:

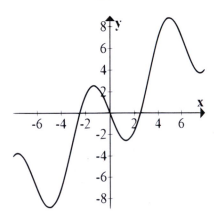

The x-intercepts of this graph are $x = -2.475$, $x = 0$, and $x = 2.475$. These are the solutions to the equation.

26. $\sin x - \cos x = x$

First we subtract x from both sides. This gives $\sin x - \cos x - x = 0$. A graph of the function $f(x) = \sin x - \cos x - x$ appears below:

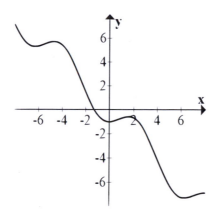

The x-intercept of this graph is $x = -1.259$. This is the solution to the equation.

28. $x^2 + 3\sin x = 0$

A graph of the function $f(x) = x^2 + 3\sin x$ appears below:

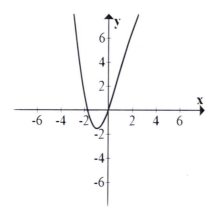

The x-intercepts of this graph are $x = -1.722$ and $x = 0$. These are the solutions to the equation.

Chapter 7

Matrices

7.1 Exercises

In exercises 1 - 10, calculate the transpose of each matrix.

In order to calculate the transpose of a matrix, interchange the rows and the columns.

2. $B = \begin{bmatrix} 2 & 1 & 5 \\ 3 & -1 & -4 \end{bmatrix}$
$\qquad B^T = \begin{bmatrix} 2 & 3 \\ 1 & -1 \\ 5 & -4 \end{bmatrix}$

6. $D = \begin{bmatrix} 5 & 4 \\ 2 & 3 \\ 0 & 1 \end{bmatrix}$
$\qquad D^T = \begin{bmatrix} 5 & 2 & 0 \\ 4 & 3 & 1 \end{bmatrix}$

10. $C = \begin{bmatrix} 1 & 2 & 3 \\ 4 & 5 & 6 \\ 7 & 8 & 9 \end{bmatrix}$
$\qquad C^T = \begin{bmatrix} 1 & 4 & 7 \\ 2 & 5 & 8 \\ 3 & 6 & 9 \end{bmatrix}$

In exercises 11 - 16, calculate the sum and difference of each pair of matrices, if possible.

12. $A = \begin{bmatrix} 2 & -1 \\ 5 & -2 \end{bmatrix}$
$\qquad B = \begin{bmatrix} -2 & 4 \\ -1 & -5 \end{bmatrix}$

$$A + B = \begin{bmatrix} 2 + (-2) & (-1) + 4 \\ 5 + (-1) & (-2) + (-5) \end{bmatrix} = \begin{bmatrix} 0 & 3 \\ 4 & -7 \end{bmatrix}$$

$$A - B = \begin{bmatrix} 2 - (-2) & (-1) - 4 \\ 5 - (-1) & (-2) - (-5) \end{bmatrix} = \begin{bmatrix} 4 & -5 \\ 6 & 3 \end{bmatrix}$$

14. $D = \begin{bmatrix} -2 & 3 \\ -5 & 4 \\ 1 & 2 \\ -6 & 3 \end{bmatrix}$ $B = \begin{bmatrix} 2 & 3 \\ 3 & -5 \\ 4 & 0 \\ 5 & 2 \end{bmatrix}$

$$D + B = \begin{bmatrix} -2+2 & 3+3 \\ -5+3 & 4+(-5) \\ 1+4 & 2+0 \\ -6+5 & 3+2 \end{bmatrix} = \begin{bmatrix} 0 & 6 \\ -2 & -1 \\ 5 & 2 \\ -1 & 5 \end{bmatrix}$$

$$D - B = \begin{bmatrix} -2-2 & 3-3 \\ -5-3 & 4-(-5) \\ 1-4 & 2-0 \\ -6-5 & 3-2 \end{bmatrix} = \begin{bmatrix} -4 & 0 \\ -8 & 9 \\ -3 & 2 \\ -11 & 1 \end{bmatrix}$$

16. $A = \begin{bmatrix} -3 & 2 \\ 2 & -1 \\ 0 & -2 \\ 2 & 3 \end{bmatrix}$ $C = \begin{bmatrix} 1 & 2 & 3 \\ 4 & 5 & 6 \\ 7 & 8 & 9 \end{bmatrix}$

These matrices cannot be added because they have different dimensions (A is 4×2 and B is 3×3).

7.2   Exercises

In exercises 1 - 10, calculate cA for the values given.

2. $c = 2, A = \begin{bmatrix} 3 & -3 \\ 7 & 2 \end{bmatrix}$

$$cA = 2 \begin{bmatrix} 3 & -3 \\ 7 & 2 \end{bmatrix} = \begin{bmatrix} 2(3) & 2(-3) \\ 2(7) & 2(2) \end{bmatrix} = \begin{bmatrix} 6 & -6 \\ 14 & 4 \end{bmatrix}$$

6. $c = -5, A = \begin{bmatrix} 1 & 2 & -3 & 2 \end{bmatrix}$

$$cA = -5 \begin{bmatrix} 1 & 2 & -3 & 2 \end{bmatrix} = \begin{bmatrix} -5(1) & -5(2) & -5(-3) & -5(2) \end{bmatrix} = \begin{bmatrix} -5 & -10 & 15 & -10 \end{bmatrix}$$

8. $c = -\dfrac{1}{3}, A = \begin{bmatrix} 3 & 6 \\ -9 & 12 \\ 15 & 0 \end{bmatrix}$

$$cA = -\frac{1}{3} \begin{bmatrix} 3 & 6 \\ -9 & 12 \\ 15 & 0 \end{bmatrix} = \begin{bmatrix} -\dfrac{3}{3} & -\dfrac{6}{3} \\ \dfrac{9}{3} & -\dfrac{12}{3} \\ -\dfrac{15}{3} & 0 \end{bmatrix} = \begin{bmatrix} -1 & -2 \\ 3 & -4 \\ -5 & 0 \end{bmatrix}$$

In exercises 11 - 20, use the following matrices:

$$A = \begin{bmatrix} -1 & 0 & 2 \\ -2 & 3 & 5 \end{bmatrix} \quad B = \begin{bmatrix} 0 & 1 & -3 \\ 2 & -4 & 5 \end{bmatrix}$$

Furthermore, $b = 2$ and $c = -3$. Calculate each of the following:

12.

$$\begin{aligned} 4A + B &= 4 \begin{bmatrix} -1 & 0 & 2 \\ -2 & 3 & 5 \end{bmatrix} + \begin{bmatrix} 0 & 1 & -3 \\ 2 & -4 & 5 \end{bmatrix} \\ &= \begin{bmatrix} 4(-1) + 0 & 4(0) + 1 & 4(2) + (-3) \\ 4(-2) + 2 & 4(3) + (-4) & 4(5) + 5 \end{bmatrix} \\ &= \begin{bmatrix} -4 & 1 & 8 \\ -6 & 8 & 25 \end{bmatrix} \end{aligned}$$

14.

$$\begin{aligned} bA + cB &= 2 \begin{bmatrix} -1 & 0 & 2 \\ -2 & 3 & 5 \end{bmatrix} - 3 \begin{bmatrix} 0 & 1 & -3 \\ 2 & -4 & 5 \end{bmatrix} \\ &= \begin{bmatrix} 2(-1) - 3(0) & 2(0) - 3(1) & 2(2) - 3(-3) \\ 2(-2) - 3(2) & 2(3) - 3(-4) & 2(5) - 3(5) \end{bmatrix} \\ &= \begin{bmatrix} -2 & -3 & 13 \\ -10 & 18 & -5 \end{bmatrix} \end{aligned}$$

20.

$$\begin{aligned} 4cA + B &= 4(-3) \begin{bmatrix} -1 & 0 & 2 \\ -2 & 3 & 5 \end{bmatrix} + \begin{bmatrix} 0 & 1 & -3 \\ 2 & -4 & 5 \end{bmatrix} \\ &= \begin{bmatrix} -12(-1) + 0 & -12(0) + 1 & -12(2) + (-3) \\ -12(-2) + 2 & -12(3) + (-4) & -12(5) + 5 \end{bmatrix} \\ &= \begin{bmatrix} 12 & 1 & -27 \\ 26 & -40 & -55 \end{bmatrix} \end{aligned}$$

In exercises 21 - 30, calculate AB and BA if possible.

In order for the matrix multiplication AB to be defined, the number of columns in A must be equal to the number of rows in B. The product matrix then has the same number of rows as A and the same number of columns as B.

22. $A = \begin{bmatrix} -1 & 0 & 2 \\ -2 & 3 & 5 \end{bmatrix}$
    $\qquad B = \begin{bmatrix} 1 & 2 \\ 2 & -4 \\ 3 & 6 \end{bmatrix}$

The dimensions of A are 2×3 and the dimensions of B are 3×2. Therefore we can calculate both AB and BA. The dimensions of AB are 2×2:

$$
\begin{aligned}
AB &= \begin{bmatrix} -1 & 0 & 2 \\ -2 & 3 & 5 \end{bmatrix} \begin{bmatrix} 1 & 2 \\ 2 & -4 \\ 3 & 6 \end{bmatrix} \\[2mm]
&= \begin{bmatrix} (-1)(1) + (0)(2) + (2)(3) & (-1)(2) + (0)(-4) + (2)(6) \\ (-2)(1) + (3)(2) + (5)(3) & (-2)(2) + (3)(-4) + (5)(6) \end{bmatrix} \\[2mm]
&= \begin{bmatrix} 5 & 10 \\ 19 & 18 \end{bmatrix}
\end{aligned}
$$

Now for BA. The dimensions of this matrix are 3×3.

$$
\begin{aligned}
BA &= \begin{bmatrix} 1 & 2 \\ 2 & -4 \\ 3 & 6 \end{bmatrix} \begin{bmatrix} -1 & 0 & 2 \\ -2 & 3 & 5 \end{bmatrix} \\[2mm]
&= \begin{bmatrix} (1)(-1) + (2)(-2) & (1)(0) + (2)(3) & (1)(2) + (2)(5) \\ (2)(-1) + (-4)(-2) & (2)(0) + (-4)(3) & (2)(2) + (-4)(5) \\ (3)(-1) + (6)(-2) & (3)(0) + (6)(3) & (3)(2) + (6)(5) \end{bmatrix} \\[2mm]
&= \begin{bmatrix} -5 & 6 & 12 \\ 6 & -12 & -16 \\ -15 & 18 & 36 \end{bmatrix}
\end{aligned}
$$

24. $A = \begin{bmatrix} -3 & 2 \\ 4 & 5 \end{bmatrix}$ $\qquad B = \begin{bmatrix} 0 & -4 \\ -6 & 5 \end{bmatrix}$

The dimensions of A are 2×2 and the dimensions of B are 2×2. Therefore both AB and BA are well defined, and both products have dimensions 2×2.

$$
\begin{aligned}
AB &= \begin{bmatrix} -3 & 2 \\ 4 & 5 \end{bmatrix} \begin{bmatrix} 0 & -4 \\ -6 & 5 \end{bmatrix} \\
&= \begin{bmatrix} (-3)(0) + (2)(-6) & (-3)(-4) + (2)(5) \\ (4)(0) + (5)(-6) & (4)(-4) + (5)(5) \end{bmatrix} \\
&= \begin{bmatrix} -12 & 22 \\ -30 & 9 \end{bmatrix}
\end{aligned}
$$

Now for BA:

$$
\begin{aligned}
BA &= \begin{bmatrix} 0 & -4 \\ -6 & 5 \end{bmatrix} \begin{bmatrix} -3 & 2 \\ 4 & 5 \end{bmatrix} \\
&= \begin{bmatrix} (0)(-3) + (-4)(4) & (0)(2) + (-4)(5) \\ (-6)(-3) + 5(4) & (-6)(2) + 5(5) \end{bmatrix} \\
&= \begin{bmatrix} -16 & -20 \\ 38 & 13 \end{bmatrix}
\end{aligned}
$$

7.3   Exercises

For each system of equations, write the equivalent augmented matrix. Then use row reduction to find the solution.

4. $\begin{cases} 2x - y &= 8 \\ x + 3y &= 11 \end{cases}$

First we write the system in augmented form:

$$\left[\begin{array}{cc|c} 2 & -1 & 8 \\ 1 & 3 & 11 \end{array}\right]$$

Now switch the rows to put a 1 in the upper left corner:

$$\left[\begin{array}{cc|c} 1 & 3 & 11 \\ 2 & -1 & 8 \end{array}\right]$$

Now add -2 times the first row to the second row:

$$\left[\begin{array}{cc|c} 1 & 3 & 11 \\ 0 & -7 & -14 \end{array}\right]$$

Now divide the second row by -7:

$$\left[\begin{array}{cc|c} 1 & 3 & 11 \\ 0 & 1 & 2 \end{array}\right]$$

Now we rewrite the system with equations instead of the matrix:

$$\begin{cases} x + 3y &= 11 \\ y &= 2 \end{cases}$$

This gives $y = 2$. Substituting this into the first equation, we get $x + 3(2) = 11$, so $x = 5$. The solution as an ordered pair is $(5, 2)$.

6. $\begin{cases} -x + 3y &= 6 \\ 3x - 9y &= 9 \end{cases}$

First we write the system in augmented form:

$$\left[\begin{array}{cc|c} -1 & 3 & 6 \\ 3 & -9 & 9 \end{array}\right]$$

Next switch rows:

$$\left[\begin{array}{cc|c} 3 & -9 & 9 \\ -1 & 3 & 6 \end{array}\right]$$

Next divide the first row by 3:

$$\left[\begin{array}{cc|c} 1 & -3 & 3 \\ -1 & 3 & 6 \end{array}\right]$$

Next add the first row to the second row:

$$\left[\begin{array}{cc|c} 1 & -3 & 3 \\ 0 & 0 & 9 \end{array}\right]$$

Now we rewrite the system in equation form:

$$\begin{cases} x - 3y & = & 3 \\ 0 & = & 9 \end{cases}$$

This system of equations has no solution.

14. $\begin{cases} 2x - 3y - z & = & 0 \\ -x + 2y + z & = & 5 \\ 3x - 4y - z & = & 1 \end{cases}$

First we write the system in augmented form:

$$\left[\begin{array}{ccc|c} 2 & -3 & -1 & 0 \\ -1 & 2 & 1 & 5 \\ 3 & -4 & -1 & 1 \end{array}\right]$$

Next switch the first and second row:

$$\left[\begin{array}{ccc|c} -1 & 2 & 1 & 5 \\ 2 & -3 & -1 & 0 \\ 3 & -4 & -1 & 1 \end{array}\right]$$

Next multiply the first row by -1:

$$\left[\begin{array}{ccc|c} 1 & -2 & -1 & -5 \\ 2 & -3 & -1 & 0 \\ 3 & -4 & -1 & 1 \end{array}\right]$$

Next add -2 times the first row to the second row:

$$\left[\begin{array}{ccc|c} 1 & -2 & -1 & -5 \\ 0 & 1 & 1 & 10 \\ 3 & -4 & -1 & 1 \end{array}\right]$$

Next add -3 times the first row to the third row:

$$\left[\begin{array}{ccc|c} 1 & -2 & -1 & -5 \\ 0 & 1 & 1 & 10 \\ 0 & 2 & 2 & 16 \end{array}\right]$$

Next add -2 times the second row to the third row:

$$\left[\begin{array}{ccc|c} 1 & -2 & -1 & -5 \\ 0 & 1 & 1 & 10 \\ 0 & 0 & 0 & -4 \end{array}\right]$$

We rewrite the system in equation form:

$$\begin{cases} x - 2y - z &= -5 \\ y + z &= 10 \\ 0 &= -4 \end{cases}$$

This system of equations has no solution.

$$20. \quad \begin{cases} x - y + z &= -4 \\ 2x - 3y + 4z &= -15 \\ 5x + y - 2z &= 12 \end{cases}$$

First we write the system in augmented form:

$$\left[\begin{array}{ccc|c} 1 & -1 & 1 & -4 \\ 2 & -3 & 4 & -15 \\ 5 & 1 & -2 & 12 \end{array} \right]$$

Next we add -2 times the first row to the second row:

$$\left[\begin{array}{ccc|c} 1 & -1 & 1 & -4 \\ 0 & -1 & 2 & -7 \\ 5 & 1 & -2 & 12 \end{array} \right]$$

Next we add -5 times the first row to the third row:

$$\left[\begin{array}{ccc|c} 1 & -1 & 1 & -4 \\ 0 & -1 & 2 & -7 \\ 0 & 6 & -7 & 32 \end{array} \right]$$

Next we add 6 times the second row to the third row:

$$\left[\begin{array}{ccc|c} 1 & -1 & 1 & -4 \\ 0 & -1 & 2 & -7 \\ 0 & 0 & 5 & -10 \end{array} \right]$$

Now multiply the second row by -1:

$$\left[\begin{array}{ccc|c} 1 & -1 & 1 & -4 \\ 0 & 1 & -2 & 7 \\ 0 & 0 & 5 & -10 \end{array} \right]$$

Finally, divide the last row by 5:

$$\left[\begin{array}{ccc|c} 1 & -1 & 1 & -4 \\ 0 & 1 & -2 & 7 \\ 0 & 0 & 1 & -2 \end{array} \right]$$

Now we rewrite the system in equation form:

$$\begin{cases} x - y + z &= -4 \\ y - 2z &= 7 \\ z &= -2 \end{cases}$$

Therefore $z = -2$. Putting this into the second equation, we get $y - 2(-2) = 7$, so $y = 3$. Putting these both into the first equation and solving, we get $x = 1$. Therefore the solution to the system is $(1, 3, -2)$.

7.4 Exercises

In exercises 1 - 10, calculate the determinant of each matrix.

Given a 2×2 matrix $A = \begin{bmatrix} a & b \\ c & d \end{bmatrix}$, the determinant of the matrix is calculated using the formula $\det A = ad - bc$.

2. $C = \begin{bmatrix} 3 & -3 \\ 7 & 2 \end{bmatrix}$

$\det C = (3)(2) - (7)(-3) = 27$

4. $B = \begin{bmatrix} 0 & 2 \\ 4 & 5 \end{bmatrix}$

$\det B = (0)(5) - (4)(2) = -8$

8. $D = \begin{bmatrix} -3 & 2 \\ 4 & 5 \end{bmatrix}$

$\det D = (-3)(5) - (4)(2) = -23$

In exercises 11 - 20, calculate the inverse of each matrix, if it exists.

Given a 2×2 matrix $A = \begin{bmatrix} a & b \\ c & d \end{bmatrix}$, the inverse of the matrix is given by the formula

$$A^{-1} = \frac{1}{\det A} \begin{bmatrix} d & -b \\ -c & b \end{bmatrix}$$

If $\det A = 0$ then the inverse does not exist.

12. $C = \begin{bmatrix} 3 & -3 \\ 7 & 2 \end{bmatrix}$

$$C^{-1} = \frac{1}{27}\begin{bmatrix} 2 & 3 \\ -7 & 3 \end{bmatrix} = \begin{bmatrix} \dfrac{2}{27} & \dfrac{1}{9} \\ -\dfrac{7}{27} & \dfrac{1}{9} \end{bmatrix}$$

14. $B = \begin{bmatrix} 0 & 2 \\ 4 & 5 \end{bmatrix}$

$$B^{-1} = \frac{1}{-8}\begin{bmatrix} 5 & -2 \\ -4 & 0 \end{bmatrix} = \begin{bmatrix} -\dfrac{5}{8} & \dfrac{1}{4} \\ \dfrac{1}{2} & 0 \end{bmatrix}$$

18. $D = \begin{bmatrix} -3 & 2 \\ 4 & 5 \end{bmatrix}$

$$D^{-1} = \frac{1}{-23}\begin{bmatrix} 5 & -2 \\ -4 & -3 \end{bmatrix} = \begin{bmatrix} -\dfrac{5}{23} & \dfrac{2}{23} \\ \dfrac{4}{23} & \dfrac{3}{23} \end{bmatrix}$$

In exercises 21 - 25 verify that matrices A and B are inverses of each other.

In order to verify that two matrices A and B are inverses, we show that $AB = BA = I$.

22. $A = \begin{bmatrix} 1 & 0 & -1 \\ 3 & -1 & 0 \\ 0 & -1 & 1 \end{bmatrix}$ $B = \frac{1}{2}\begin{bmatrix} -1 & 1 & -1 \\ -3 & 1 & -3 \\ -3 & 1 & -1 \end{bmatrix}$

$$\begin{aligned}
AB &= \begin{bmatrix} 1 & 0 & -1 \\ 3 & -1 & 0 \\ 0 & -1 & 1 \end{bmatrix}\left(\frac{1}{2}\begin{bmatrix} -1 & 1 & -1 \\ -3 & 1 & -3 \\ -3 & 1 & -1 \end{bmatrix}\right) \\
&= \frac{1}{2}\begin{bmatrix} (1)(-1)+0+(-1)(-3) & (1)(1)+0+(-1)(1) & (1)(-1)+0+(-1)(-1) \\ (3)(-2)+(-1)(-3)+0 & (3)(1)+(-1)(1)+0 & (3)(-1)+(-1)(-3)+0 \\ 0+(-1)(-3)+(1)(-3) & 0+(-1)(1)+(1)(1) & 0+(-1)(-3)+(1)(-2) \end{bmatrix} \\
&= \frac{1}{2}\begin{bmatrix} 1 & 0 & 0 \\ 0 & 1 & 0 \\ 0 & 0 & 1 \end{bmatrix} \\
&= I_2
\end{aligned}$$

$$BA = \frac{1}{2} \begin{bmatrix} -1 & 1 & -1 \\ -3 & 1 & -3 \\ -3 & 1 & -1 \end{bmatrix} \begin{bmatrix} 1 & 0 & -1 \\ 3 & -1 & 0 \\ 0 & -1 & 1 \end{bmatrix}$$

$$= \frac{1}{2} \begin{bmatrix} (-1)(1)+(1)(3)+0 & 0+(1)(-1)+(-1)(-1) & (-1)(-1)+0+(-1)(1) \\ (-3)(1)+(1)(3)+0 & 0+(1)(-1)+(-3)(-1) & (-3)(-1)+0+(-3)(1) \\ (-3)(1)+(1)(3)+0 & 0+(1)(-1)+(-1)(-1) & (-3)(-1)+0+(-1)(1) \end{bmatrix}$$

$$= \frac{1}{2} \begin{bmatrix} 2 & 0 & 0 \\ 0 & 2 & 0 \\ 0 & 0 & 2 \end{bmatrix}$$

$$= I_2$$

7.5 Exercises

Calculate the determinant of each of the following matrices:

2. $B = \begin{bmatrix} 3 & -1 \\ 2 & 1 \end{bmatrix}$

det $B = (3)(1) - (-1)(2) = 5$

4. $D = \begin{bmatrix} 1 & 2 \\ 2 & 4 \end{bmatrix}$

det $D = (1)(4) - (2)(2) = 0$

8. $D = \begin{bmatrix} 2 & -1 & 3 \\ 3 & 1 & -2 \\ 1 & -3 & 2 \end{bmatrix}$

$$\begin{aligned} \det D &= 2\det \begin{bmatrix} 1 & -2 \\ -3 & 2 \end{bmatrix} - (-1)\det \begin{bmatrix} 3 & -2 \\ 1 & 2 \end{bmatrix} + 3\det \begin{bmatrix} 3 & 1 \\ 1 & -3 \end{bmatrix} \\ &= 2((1)(2)-(-2)(-3)) + ((3)(2)-(1)(-2)) + 3((3)(-3)-(1)(1)) \\ &= 2(-4) + 8 + 3(-10) \\ &= -30 \end{aligned}$$

12. $D = \begin{bmatrix} 3 \\ -1 \end{bmatrix}$

It is not possible to calculate the determinant of this matrix, because it does not have the same number of rows and columns.

Calculate the inverse of each of the following matrices:

18. $B = \begin{bmatrix} 3 & -1 \\ 2 & 1 \end{bmatrix}$

$$B^{-1} = \frac{1}{(3)(1) - (-1)(2)} \begin{bmatrix} 1 & 1 \\ -2 & 3 \end{bmatrix} = \begin{bmatrix} \dfrac{1}{5} & \dfrac{1}{5} \\ -\dfrac{2}{5} & \dfrac{3}{5} \end{bmatrix}$$

20. $D = \begin{bmatrix} 1 & 2 \\ 2 & 4 \end{bmatrix}$

This inverse does not exist because the determinant of D is zero.

22. $B = \begin{bmatrix} 0 & 1 & 0 \\ 1 & 0 & 1 \\ 0 & 1 & 0 \end{bmatrix}$

For this problem we will use an augmented matrix and row reduction.

$$\left[\begin{array}{ccc|ccc} 0 & 1 & 0 & 1 & 0 & 0 \\ 1 & 0 & 1 & 0 & 1 & 0 \\ 0 & 1 & 0 & 0 & 0 & 1 \end{array}\right]$$

First we will add row two to row one in order to make the first entry in row one nonzero:

$$\left[\begin{array}{ccc|ccc} 1 & 1 & 1 & 1 & 1 & 0 \\ 1 & 0 & 1 & 0 & 1 & 0 \\ 0 & 1 & 0 & 0 & 0 & 1 \end{array}\right]$$

Now subtract row one from row two:

$$\left[\begin{array}{ccc|ccc} 1 & 1 & 1 & 1 & 1 & 0 \\ 0 & -1 & 0 & -1 & 0 & 0 \\ 0 & 1 & 0 & 0 & 0 & 1 \end{array}\right]$$

Next add row two to row one:

$$\left[\begin{array}{ccc|ccc} 1 & 0 & 1 & 0 & 1 & 0 \\ 0 & -1 & 0 & -1 & 0 & 0 \\ 0 & 1 & 0 & 0 & 0 & 1 \end{array}\right]$$

Now add row two to row three:

$$\left[\begin{array}{ccc|ccc} 1 & 0 & 1 & 0 & 1 & 0 \\ 0 & -1 & 0 & -1 & 0 & 0 \\ 0 & 0 & 0 & -1 & 0 & 1 \end{array}\right]$$

Since the first three entries in the third row are zero, this matrix does not have an inverse.

24. $D = \begin{bmatrix} 2 & -1 & 3 \\ 3 & 1 & -2 \\ 1 & -3 & 2 \end{bmatrix}$

First we write D with the identity matrix:

$$\begin{bmatrix} 2 & -1 & 3 & | & 1 & 0 & 0 \\ 3 & 1 & -2 & | & 0 & 1 & 0 \\ 1 & -3 & 2 & | & 0 & 0 & 1 \end{bmatrix}$$

We add $-\dfrac{3}{2}$ times the first row to the second row:

$$\begin{bmatrix} 2 & -1 & 3 & | & 1 & 0 & 0 \\ 0 & \dfrac{5}{2} & -\dfrac{13}{2} & | & -\dfrac{3}{2} & 1 & 0 \\ 1 & -3 & 2 & | & 0 & 0 & 1 \end{bmatrix}$$

Now we add $-\dfrac{1}{2}$ times the first row to the third row:

$$\begin{bmatrix} 2 & -1 & 3 & | & 1 & 0 & 0 \\ 0 & \dfrac{5}{2} & -\dfrac{13}{2} & | & -\dfrac{3}{2} & 1 & 0 \\ 0 & -\dfrac{5}{2} & \dfrac{1}{2} & | & -\dfrac{1}{2} & 0 & 1 \end{bmatrix}$$

Now double the last two rows:

$$\begin{bmatrix} 2 & -1 & 3 & | & 1 & 0 & 0 \\ 0 & 5 & -13 & | & -3 & 2 & 0 \\ 0 & -5 & 1 & | & -1 & 0 & 2 \end{bmatrix}$$

Next add $\dfrac{1}{5}$ times the second row to the first row:

$$\begin{bmatrix} 2 & 0 & \dfrac{2}{5} & | & \dfrac{2}{5} & \dfrac{2}{5} & 0 \\ 0 & 5 & -13 & | & -3 & 2 & 0 \\ 0 & -5 & 1 & | & -1 & 0 & 2 \end{bmatrix}$$

Now add the second row to the third row:

$$\begin{bmatrix} 2 & 0 & \dfrac{2}{5} & | & \dfrac{2}{5} & \dfrac{2}{5} & 0 \\ 0 & 5 & -13 & | & -3 & 2 & 0 \\ 0 & 0 & -12 & | & -4 & 2 & 2 \end{bmatrix}$$

Now divide the first row by 2, the second row by 5, and the third row by -12:

$$\left[\begin{array}{ccc|ccc} 1 & 0 & \dfrac{1}{5} & \dfrac{1}{5} & \dfrac{1}{5} & 0 \\[3mm] 0 & 1 & -\dfrac{13}{5} & -\dfrac{3}{5} & \dfrac{2}{5} & 0 \\[3mm] 0 & 0 & 1 & \dfrac{1}{3} & -\dfrac{1}{6} & -\dfrac{1}{6} \end{array}\right]$$

Next add $-\dfrac{1}{5}$ times the third row to the first row:

$$\left[\begin{array}{ccc|ccc} 1 & 0 & 0 & \dfrac{2}{15} & \dfrac{7}{30} & \dfrac{1}{30} \\[3mm] 0 & 1 & -\dfrac{13}{5} & -\dfrac{3}{5} & \dfrac{2}{5} & 0 \\[3mm] 0 & 0 & 1 & \dfrac{1}{3} & -\dfrac{1}{6} & -\dfrac{1}{6} \end{array}\right]$$

Finally, add $\dfrac{13}{5}$ times the third row to the second row:

$$\left[\begin{array}{ccc|ccc} 1 & 0 & 0 & \dfrac{2}{15} & \dfrac{7}{30} & \dfrac{1}{30} \\[3mm] 0 & 1 & 0 & \dfrac{4}{15} & -\dfrac{1}{30} & -\dfrac{13}{30} \\[3mm] 0 & 0 & 1 & \dfrac{1}{3} & -\dfrac{1}{6} & -\dfrac{1}{6} \end{array}\right]$$

We have the identity matrix on the left of the vertical line. The right side is the inverse.

$$\text{Therefore } D^{-1} = \left[\begin{array}{ccc} \dfrac{2}{15} & \dfrac{7}{30} & \dfrac{1}{30} \\[3mm] \dfrac{4}{15} & -\dfrac{1}{30} & -\dfrac{13}{30} \\[3mm] \dfrac{1}{3} & -\dfrac{1}{6} & -\dfrac{1}{6} \end{array}\right]$$

7.6 Exercises

For each system of equations in problems 1 - 10, identify the coefficient matrix A, the variable matrix X, and the constant matrix B.

2. $\begin{cases} 4x + 3y &= 8 \\ 2x + 2y &= 6 \end{cases}$

$A = \begin{bmatrix} 4 & 3 \\ 2 & 2 \end{bmatrix} \qquad X = \begin{bmatrix} x \\ y \end{bmatrix} \qquad B = \begin{bmatrix} 8 \\ 6 \end{bmatrix}$

6. $\begin{cases} -x + 3y &= 6 \\ 3x - 9y &= 9 \end{cases}$

$A = \begin{bmatrix} -1 & 3 \\ 3 & -9 \end{bmatrix} \qquad X = \begin{bmatrix} x \\ y \end{bmatrix} \qquad B = \begin{bmatrix} 6 \\ 9 \end{bmatrix}$

10. $\begin{cases} 2x - 3y &= 0 \\ 2x + 6y &= 3 \end{cases}$

$A = \begin{bmatrix} 2 & -3 \\ 2 & 6 \end{bmatrix} \qquad X = \begin{bmatrix} x \\ y \end{bmatrix} \qquad B = \begin{bmatrix} 0 \\ 3 \end{bmatrix}$

For problems 11 - 20, calculate the inverse of the coefficient matrix. Then use this to solve the system using the formula $X = A^{-1}B$, if possible. If the inverse of the coefficient matrix does not exist, then solve the system using alternate methods. These are the same systems as problems 1 - 10.

12. $\begin{cases} 4x + 3y &= 8 \\ 2x + 2y &= 6 \end{cases}$

Since $A = \begin{bmatrix} 4 & 3 \\ 2 & 2 \end{bmatrix}$, we have $A^{-1} = \dfrac{1}{(4)(2) - (3)(2)} \begin{bmatrix} 2 & -3 \\ -2 & 4 \end{bmatrix} = \begin{bmatrix} 1 & -\frac{3}{2} \\ -1 & 2 \end{bmatrix}$

Therefore $X = A^{-1}B = \begin{bmatrix} 1 & -\frac{3}{2} \\ -1 & 2 \end{bmatrix} \begin{bmatrix} 8 \\ 6 \end{bmatrix} = \begin{bmatrix} -1 \\ 4 \end{bmatrix}$

We have $x = -1$ and $y = 4$.

16. $\begin{cases} -x + 3y &= 6 \\ 3x - 9y &= 9 \end{cases}$

Since $A = \begin{bmatrix} -1 & 3 \\ 3 & -9 \end{bmatrix}$, we have $\det A = (-1)(-9) - (3)(3) = 0$. Since the determinant of A

is zero, we need to use another method.

$$\begin{bmatrix} -1 & 3 & | & 6 \\ 3 & -9 & | & 9 \end{bmatrix}$$

We add 3 times the first row to the second row:

$$\begin{bmatrix} -1 & 3 & | & 6 \\ 0 & 0 & | & 27 \end{bmatrix}$$

The system in equation form is

$$\begin{cases} -x + 37 & = & 6 \\ 0 & = & 27 \end{cases}$$

This system has no solutions.

20. $\begin{cases} 2x - 3y & = & 0 \\ 2x + 6y & = & 3 \end{cases}$

Since $A = \begin{bmatrix} 2 & -3 \\ 2 & 6 \end{bmatrix}$, we have $A^{-1} = \dfrac{1}{(2)(6) - (2)(-3)} \begin{bmatrix} 6 & 3 \\ -2 & 2 \end{bmatrix} = \begin{bmatrix} \dfrac{1}{3} & \dfrac{1}{6} \\ -\dfrac{1}{9} & \dfrac{1}{9} \end{bmatrix}$

Therefore $X = A^{-1}B = \begin{bmatrix} \dfrac{1}{3} & \dfrac{1}{6} \\ -\dfrac{1}{9} & \dfrac{1}{9} \end{bmatrix} \begin{bmatrix} 0 \\ 3 \end{bmatrix} = \begin{bmatrix} \dfrac{1}{2} \\ \dfrac{1}{3} \end{bmatrix}.$

This gives $x = \dfrac{1}{2}$ and $y = \dfrac{1}{3}$.

7.7   Exercises

Use Cramer's Rule to solve each system of equations.

2. $\begin{cases} 4x + 3y & = & 8 \\ 2x + 2y & = & 6 \end{cases}$

We have $A = \begin{bmatrix} 4 & 3 \\ 2 & 2 \end{bmatrix}$, $A_1 = \begin{bmatrix} 8 & 3 \\ 6 & 2 \end{bmatrix}$, and $A_2 = \begin{bmatrix} 4 & 8 \\ 2 & 6 \end{bmatrix}$.

Therefore $\det A = (4)(2) - (3)(2) = 2$, $\det A_1 = (8)(2) - (3)(6) = -2$, and

$\det A_2 = (4)(6) - (2)(8) = 8$.

This gives $x = \dfrac{\det A_1}{\det A} = \dfrac{-2}{2} = -1$, $y = \dfrac{\det A_2}{\det A} = \dfrac{8}{2} = 4$.

4. $\begin{cases} 2x - y = 8 \\ x + 3y = 11 \end{cases}$

We have $A = \begin{bmatrix} 2 & -1 \\ 1 & 3 \end{bmatrix}$, $A_1 = \begin{bmatrix} 8 & -1 \\ 11 & 3 \end{bmatrix}$, $A_2 = \begin{bmatrix} 2 & 8 \\ 1 & 11 \end{bmatrix}$.

Therefore $\det A = (2)(3) - (1)(-1) = 7$, $\det A_1 = (8)(3) - (-1)(11) = 35$, and

$\det A_2 = (2)(11) - (1)(8) = 14$

This gives $x = \dfrac{\det A_1}{\det A} = \dfrac{35}{7} = 5$, $y = \dfrac{\det A_2}{\det A} = \dfrac{14}{7} = 2$.

12. $\begin{cases} 2x + y - 3z = 0 \\ -2x + 2y + z = -7 \\ 3x - 4y - 3z = 7 \end{cases}$

We have $A = \begin{bmatrix} 2 & 1 & -3 \\ -2 & 2 & 1 \\ 3 & -4 & -3 \end{bmatrix}$, $A_1 = \begin{bmatrix} 0 & 1 & -3 \\ -7 & 2 & 1 \\ 7 & -4 & -3 \end{bmatrix}$, $A_2 = \begin{bmatrix} 2 & 0 & -3 \\ -2 & -7 & 1 \\ 3 & 7 & -3 \end{bmatrix}$, and

$A_3 = \begin{bmatrix} 2 & 1 & 0 \\ -2 & 2 & -7 \\ 3 & -4 & 7 \end{bmatrix}$

Therefore

$\det A = 2((2)(-3) - (1)(-4)) - (1)((-2)(-3) - (1)(3)) + (-3)((-2)(-4) - (2)(3)) = -13$

$\det A_1 = 0 - (1)((-7)(-3) - (1)(7)) + (-3)((-7)(-4) - (2)(3)) = -80$

$\det A_2 = 2((2)(7) - (-7)(-4)) - 0 + (-3)((-2)(7) - (-7)(3)) = -105$

$\det A_3 = 2((2)(7) - (-4)(-7)) - (1)((-2)(7) - (-7)(3)) + 0 = -35$

$x = \dfrac{\det A_1}{\det A} = \dfrac{80}{13}$, $y = \dfrac{\det A_2}{\det A} = \dfrac{105}{13}$, $z = \dfrac{\det A_3}{\det A} = \dfrac{35}{13}$

7.8 Exercises

In problems 1 - 10, let $\mathbf{u} = \langle 1, 4 \rangle$ *and* $\mathbf{v} = \langle 3, -5 \rangle$*. Calculate each of the following:*

In order to add two vectors $\mathbf{u} = \langle u_1, u_2 \rangle$ and $\mathbf{v} = \langle v_1, v_2 \rangle$, we add the corresponding components: $\mathbf{u}+\mathbf{v} = \langle u_1+v_1, u_2+v_2 \rangle$. To multiply a vector by a scalar c, multiply each component by c: $c\mathbf{u} = c\langle u_1, u_2 \rangle = \langle cu_1, cu_2 \rangle$. The magnitude (length) of the vector $\mathbf{u}$ is given by the formula $|\mathbf{u} = \sqrt{u_1^2 + u_2^2}$.

2. $-3\mathbf{u} = -3\langle 1, 4 \rangle = \langle -3, -12 \rangle$

4. $-4\mathbf{v} = -4\langle 3, -5 \rangle = \langle -12, 20 \rangle$

6. $4\mathbf{u} - 3\mathbf{v} = 4\langle 1, 4 \rangle - 3\langle 3, -5 \rangle = \langle 4, 16 \rangle - \langle 9, -15 \rangle = \langle 4 - 9, 16 - (-15) \rangle = \langle -5, 31 \rangle$

8. $|\mathbf{v}| = \sqrt{3^2 + (-5)^2} = \sqrt{9 + 25} = \sqrt{34}$

10. $|3\mathbf{u} + 4\mathbf{v}| = |3\langle 1, 4 \rangle + 4\langle 3, -5 \rangle| = |\langle 3, 12 \rangle + \langle 12, -20 \rangle| = |\langle 15, -8 \rangle| = \sqrt{15^2 + (-8)^2} = \sqrt{225 + 64} = 17$

In problems 11 - 20, find a vector with initial point A and terminal point B, and draw a sketch of the vector in standard position.

Given a point A with coordinates (x_1, y_1) and B with coordinates (x_2, y_2), the vector that starts at point A and ends at point B is given by the formula $\vec{AB} = \langle x_2 - x_1, y_2 - y_1 \rangle$.

12. $A = (2, 5)$ $B = (3, 8)$

$\vec{AB} = \langle 8 - 5, 3 - 2 \rangle = \langle 3, 1 \rangle$

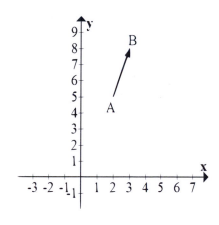

18. $A = (-6, 2)$ $B = (-4, -3)$

$\vec{AB} = \langle -4 - (-6), -3 - 2 \rangle = \langle 2, -5 \rangle$

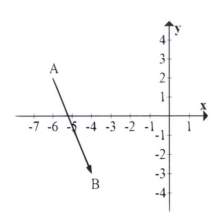

In problems 21 - 30, let the vector $\mathbf{i} = \langle 1, 0 \rangle$ *and* $\mathbf{j} = \langle 0, 1 \rangle$. *Write each of the following vectors as a linear combination of* $\mathbf{i}$ *and* $\mathbf{j}$.

Given two vectors $\mathbf{u}$ and $\mathbf{v}$ along with two scalars a and b, a linear combination of these vectors can be written in the form $a\mathbf{u} + b\mathbf{v}$. If $\mathbf{u} = \mathbf{i}$ and $\mathbf{v} = \mathbf{j}$, then $a\mathbf{i} + b\mathbf{j} = \langle a, b \rangle$. The vectors $\mathbf{i}$ and $\mathbf{j}$ are called unit component vectors.

22. $\mathbf{u} = \langle -3, 2 \rangle = -3\mathbf{i} + 2\mathbf{j}$

30. $\mathbf{v} = \langle -6, 6 \rangle = -6\mathbf{i} + 6\mathbf{j}$

In problems 31 - 40, find unit vectors in the direction of the given vector.

To find a unit vector that points in the same direction of a vector, divide both components by the magnitude of the vector.

32. $\mathbf{v} = \langle -4, 3 \rangle$

The magnitude of $\mathbf{v}$ is $|\mathbf{v}| = \sqrt{(-4)^2 + 3^2} = 5$. Dividing both components of $\mathbf{v}$ by 5, we get the unit vector $\left\langle -\dfrac{4}{5}, \dfrac{3}{5} \right\rangle$.

38. $\mathbf{v} = \langle -24, 7 \rangle$

The magnitude of $\mathbf{v}$ is $|\mathbf{v}| = \sqrt{(-24)^2 + 7^2} = 25$. Dividing both components of $\mathbf{v}$ by 25, we get the unit vector $\left\langle -\dfrac{24}{25}, \dfrac{7}{25} \right\rangle$.

42. Find a vector of magnitude 5 in the direction of $\mathbf{v} = \dfrac{5}{13}\mathbf{i} - \dfrac{12}{13}\mathbf{j}$.

First we find a unit vector in the direction of $\mathbf{v}$. Since the magnitude of $\mathbf{v}$ is $|\mathbf{v}| = \sqrt{\left(\dfrac{5}{13}\right)^2 + \left(\dfrac{12}{13}\right)^2} = \sqrt{\dfrac{25}{169} + \dfrac{144}{169}} = 1$, this vector is already a unit vector. Next we multiply this vector by 3, getting $3\mathbf{v} = \left\langle \dfrac{15}{13}, -\dfrac{36}{13} \right\rangle$. This is the desired vector.

44. Find a vector of magnitude 6 in the direction opposite to the direction of $\mathbf{u} = -12\mathbf{i} - 9\mathbf{j}$.

First we calculate the magnitude of $\mathbf{u}$: $|\mathbf{u}| = \sqrt{(-12)^2 + (-9)^2} = \sqrt{144 + 81} = \sqrt{225} = 15$. Next we divide the vector by its magnitude. This gives the new vector $\left\langle -\dfrac{4}{5}, -\dfrac{3}{5} \right\rangle$. Finally we multiply the new vector by 6, getting the final answer: $\left\langle -\dfrac{24}{5}, -\dfrac{18}{5} \right\rangle$.

46. Let $\mathbf{u} = <6, 2>$, $\mathbf{v} = <5, 3>$, and $\mathbf{w} = <18, 12>$. Find scalars a and b such that $\mathbf{w} = a\mathbf{u} + b\mathbf{v}$.

We want to solve the equation $\mathbf{w} = a\mathbf{u} + b\mathbf{v}$. Substituting the three vectors into the equation, we get $\langle 18, 12 \rangle = a\langle 6, 2 \rangle + b\langle 5, 3 \rangle$. Simplifying the right hand side, we get $\langle 18, 12 \rangle = \langle 6a + 5b, 2a + 3b \rangle$. Setting corresponding components equal to each other, we get the system of equations

$$\begin{cases} 6a + 5b &= 18 \\ 2a + 3b &= 12 \end{cases}$$

If we subtract 3 times the second equation from the first equation, we get the equation $-4b = -18$, so $b = \dfrac{9}{2}$. Solving for a, we get $a = -\dfrac{3}{4}$.

48. An airplane is flying in the direction 30° west of south at 250 mph. Find the component form of the velocity of the airplane, assuming that the positive x-axis represents due east and the positive y-axis represents due north.

A graph of the velocity vector for the plane appears below:

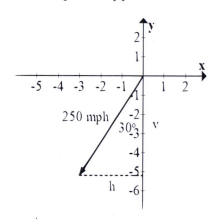

We have a right triangle with hypotenuse 250. The opposite side to the 30 degree angle is the horizontal component of the velocity, and the adjacent side is the vertical component. Therefore

$$
\begin{aligned}
\sin 30 &= \frac{h}{250} \\
h &= 250 \sin 30 \\
&= 125 \\
\cos 30 &= \frac{v}{250} \\
v &= 250 \cos 30 \\
&= 125\sqrt{3}
\end{aligned}
$$

The velocity vector in component form is $\langle -125, -125\sqrt{3} \rangle$.

50. The angle between vectors $\mathbf{u} = \langle u_1, u_2 \rangle$ and $\mathbf{v} = \langle v_1, v_2 \rangle$ can be calculated using the following formula:

$$
\cos \theta = \frac{u_1 v_1 + u_2 v_2}{|\mathbf{u}|\,|\mathbf{v}|}
$$

Calculate the angle between each pair of vectors given below:

a. $\mathbf{u} = \langle 2, 0 \rangle \quad \mathbf{v} = \langle \sqrt{3}, 1 \rangle$

$$
\begin{aligned}
\cos \theta &= \frac{u_1 v_1 + u_2 v_2}{|\mathbf{u}|\,|\mathbf{v}|} \\
&= \frac{2\sqrt{3} + (0)(1)}{\sqrt{2^2 + 0^2}\,\sqrt{(\sqrt{3})^2 + 1^2}} \\
&= \frac{2\sqrt{3}}{\sqrt{4}\,\sqrt{4}} \\
&= \frac{\sqrt{3}}{2} \\
\theta &= \frac{\pi^R}{6}
\end{aligned}
$$

b. $\mathbf{u} = \langle 2, 4 \rangle$ $\mathbf{v} = \langle 5, -2 \rangle$

$$
\begin{aligned}
\cos \theta &= \frac{u_1 v_1 + u_2 v_2}{|\mathbf{u}| \, |\mathbf{v}|} \\
&= \frac{(2)(5) + (4)(-2)}{\sqrt{2^2 + 4^2} \, \sqrt{5^2 + (-2)^2}} \\
&= \frac{2}{\sqrt{20} \, \sqrt{29}} \\
&= \frac{2}{2\sqrt{5} \, \sqrt{29}} \\
&= \frac{1}{\sqrt{145}} \\
\theta &= \arccos \left(\frac{1}{\sqrt{145}} \right) \\
&\approx 1.49^R
\end{aligned}
$$

7.9   Exercises

In problems 1 - 10, express each of the following as the sum or difference of partial fractions:

4. $\dfrac{x - 3}{x^2 + 5x + 6}$

First factor the denominator as $(x + 2)(x + 3)$. These factors become the denominators in the partial fractions:

$$
\frac{x - 3}{x^2 + 5x + 6} = \frac{A}{x + 2} + \frac{B}{x + 3}
$$

Multiplying the equation by $(x + 2)(x + 3)$, we get

$$
\begin{aligned}
(x + 2)(x + 3) \left(\frac{x - 3}{x^2 + 5x + 6} \right) &= (x + 2)(x + 3) \left(\frac{A}{x + 2} + \frac{B}{x + 3} \right) \\
x - 3 &= A(x + 3) + B(x + 2)
\end{aligned}
$$

This equation must hold true for all values of x. Substituting $x = -2$, we get

$$
\begin{aligned}
-2 - 3 &= A(-2 + 3) + B(-2 + 2) \\
A &= -5
\end{aligned}
$$

Substituting $x = -3$, we get

$$
\begin{aligned}
-3 - 3 &= A(-3 + 3) + B(-3 + 2) \\
-B &= -6 \\
B &= 6
\end{aligned}
$$

Therefore the correct solution is $\dfrac{x-3}{x^2+5x+6} = \dfrac{6}{x+3} - \dfrac{5}{x+2}$.

6. $\dfrac{x+3}{2x^3-8x}$

The denominator factors as $2x(x^2-4) = 2x(x-2)(x+2)$. This time we have three fractions.

$$\frac{x+3}{2x^3-8x} = \frac{A}{2x} + \frac{B}{x-2} + \frac{C}{x+2}$$

Multiplying both sides of the equation by the common denominator, we get

$$2x(x-2)(x+2)\left(\frac{x+3}{2x^3-8x}\right) = 2x(x-2)(x+2)\left(\frac{A}{2x} + \frac{B}{x-2} + \frac{C}{x+2}\right)$$
$$x+3 = A(x-2)(x+2) + B(2x)(x+2) + C(2x)(x-2)$$

First we set $x=0$:

$$0+3 = A(0-2)(0+2) + B(2)(0)(0+2) + C(2)(0)(0-2)$$
$$-4A = 3$$
$$A = -\frac{3}{4}$$

Next set $x=2$:

$$2+3 = A(2-2)(2+2) + B(2)(2)(2+2) + C(2)(2)(2-2)$$
$$16B = 5$$
$$B = \frac{5}{16}$$

Now let $x=-2$:

$$-2+3 = A(-2-2)(-2+2) + B(2)(-2)(-2+2) + C(2)(-2)(-2-2)$$
$$16C = 1$$
$$C = \frac{1}{16}$$

Therefore the solution is $\dfrac{x+3}{2x^3-8x} = \dfrac{-\frac{3}{4}}{2x} + \dfrac{\frac{5}{16}}{x-2} + \dfrac{\frac{1}{16}}{x+2} = -\dfrac{3}{8x} + \dfrac{5}{16(x-2)} + \dfrac{1}{16(x+2)}$.

8. $\dfrac{1}{(x+1)(x^2+1)}$

This denominator cannot be factored any further, so

$$\frac{1}{(x+1)(x^2+1)} = \frac{A}{x+1} + \frac{Bx+C}{x^2+1}$$

Multiplying both sides by the common denominator:

$$(x+1)(x^2+1)\left(\frac{1}{(x+1)(x^2+1)}\right) = (x+1)(x^2+1)\left(\frac{A}{x+1}+\frac{Bx+C}{x^2+1}\right)$$
$$1 = A(x^2+1)+(Bx+C)(x+1)$$

Setting $x = -1$, we get

$$1 = A((-1)^2+1)+(B(-1)+C)(-1+1)$$
$$2A = 1$$
$$A = \frac{1}{2}$$

Setting $x = 0$ and $A = \frac{1}{2}$, we get

$$1 = \frac{1}{2}(0^2+1)+(B(0)+C)(0+1)$$
$$1 = \frac{1}{2}+C$$
$$C = \frac{1}{2}$$

Setting $x = 1$, $A = \frac{1}{2}$, and $C = \frac{1}{2}$, we get

$$1 = \frac{1}{2}(1^2+1)+\left(B(1)+\frac{1}{2}\right)(1+1)$$
$$1 = 1+2\left(B+\frac{1}{2}\right)$$
$$1 = 1+2B+1$$
$$2B = -1$$
$$B = -\frac{1}{2}$$

Therefore the solution is $\dfrac{1}{(x+1)(x^2+1)} = \dfrac{\frac{1}{2}}{x+1}+\dfrac{-\frac{1}{2}x+\frac{1}{2}}{x^2+1} = \dfrac{1}{2}\left(\dfrac{1}{x+1}-\dfrac{x-1}{x^2+1}\right).$

In problems 11 - 18, perform long division first to write the rational function as the sum of a polynomial and rational function. Then express the rational function as the sum or difference of partial fractions:

18. $\dfrac{4x^3-x^2+3x}{x^2-3x+2}$

First we perform long division. This gives $\dfrac{4x^3-x^2+3x}{x^2-3x+2} = 4x+11+\dfrac{36x-22}{x^2-3x+2}$. Next we factor the denominator, and work only with this fraction:

$$\frac{36x-22}{x^2-3x+2} = \frac{A}{x-2}+\frac{B}{x-1}$$

Now multiply both sides by the common denominator:

$$(x-2)(x-1)\left(\frac{36x-22}{x^2-3x+2}\right) = (x-2)(x-1)\left(\frac{A}{x-2}+\frac{B}{x-1}\right)$$
$$36x-22 = A(x-1)+B(x-2)$$

Setting $x = 2$, we get

$$36(2)-22 = A(2-1)+B(2-2)$$
$$A = 50$$

Setting $x = 1$, we get

$$36(1)-22 = A(1-1)+B(1-2)$$
$$-B = 14$$
$$B = -14$$

Therefore the original function can be rewritten as $\dfrac{4x^3-x^2+3x}{x^2-3x+2}=4x+11+\dfrac{50}{x-2}-\dfrac{14}{x-1}$.

Express each of the following as a sum or difference of partial fractions:

20. $\dfrac{\sin\theta}{\sin^2\theta+\sin\theta-6}$

First we factor the denominator as $(\sin\theta+3)(\sin\theta-2)$. Therefore the decomposition takes the form

$$\frac{\sin\theta}{\sin^2\theta+\sin\theta-6} = \frac{A}{\sin\theta+3}+\frac{B}{\sin\theta-2}$$

Multiplying both side of the equation by the common denominator, we get

$$(\sin\theta+3)(\sin\theta-2)\left(\frac{\sin\theta}{\sin^2\theta+\sin\theta-6}\right) = (\sin\theta+3)(\sin\theta-2)\left(\frac{A}{\sin\theta+3}+\frac{B}{\sin\theta-2}\right)$$
$$\sin\theta = A(\sin\theta-2)+B(\sin\theta+3)$$

Setting $\theta=0$ yields the equation $\sin 0 = A(\sin 0-2)+B(\sin 0+3)$, so $0=-2A+3B$.

Setting $\theta=\dfrac{\pi}{2}$ yields the equation $\sin\dfrac{\pi}{2} = A\left(\sin\dfrac{\pi}{2}-2\right)+B\left(\sin\dfrac{\pi}{2}+3\right)$ so $1=-A+4B$.

Subtracting twice the second equation from the first equation, we get $-2=-5B$, so $B=\dfrac{2}{5}$.

Substituting this into the first equation yields $A=\dfrac{3}{5}$. Therefore the decomposition becomes

$$\frac{\sin\theta}{\sin^2\theta+\sin\theta-6} = \frac{\frac{3}{5}}{\sin\theta+3}+\frac{\frac{2}{5}}{\sin\theta-2} = \frac{1}{5}\left(\frac{3}{\sin\theta+3}+\frac{2}{\sin\theta-2}\right).$$

Chapter 8

Conic Sections

8.2   Exercises

In problems 1-10, identify the center and radius of each circle. Create a graph.

The standard for for the equation of a circle with radius r and center located at (h, k) is

$$(x - h)^2 + (y - k)^2 = r^2$$

2. $(x - 3)^2 + (x - 4)^2 = 16$

Center is at (3, 4) and radius is 4.

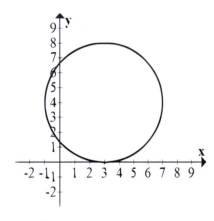

4. $(x-3)^2 + (y+2)^2 = 9$

Center is at (3, -2) and radius is 3.

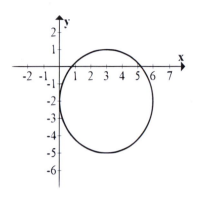

In problems 11 - 20, put each circle in standard form. Identify the center and radius, and graph.

12. $x^2 - 6x + y^2 - 4y = 12$

First we group the x terms and the y terms:

$(x^2 - 6x) + (y^2 - 4y) = 12$

Next we take half the coefficient of x and square it. Add this to both sides:

$(x^2 - 6x + 9) + (y^2 - 4y) = 12 + 9$

Now do the same for y:

$(x^2 - 6x + 9) + (y^2 - 4y + 4) = 21 + 4$

Factor both trinomials:

$(x-3)^2 + (y-2)^2 = 25$

The center is located at $(3, 2)$ and the radius is 5:

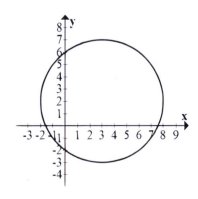

16. $x^2 + 8x + y^2 - 4y - 16 = 0$

First we group the x terms and the y terms, and move the constant to the right hand side:

$(x^2 + 8x) + (y^2 - 4y) = 16$

Next we take half the coefficient of x and square it. Add this to both sides:

$(x^2 + 8x + 16) + (y^2 - 4y) = 16 + 16$

Now do the same for y:

$(x^2 + 8x + 16) + (y^2 - 4y + 4) = 32 + 4$

Factor both trinomials:

$(x + 4)^2 + (y - 2)^2 = 36$

The center is located at $(-4, 2)$ and the radius is 6:

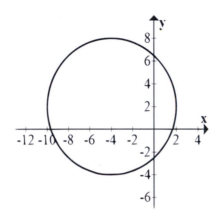

22. Write the equation of the circle whose center is the point $(4, -3)$ and whose radius is 5.

We have $h = 4$, $k = -3$, and $r = 5$. Substituting these into the standard form of a circle yields $(x - 4)^2 + (y + 3)^2 = 25$.

24. A circle has a diameter whose end points are $(-3, 2)$ and $(5, 6)$. Find its equation.

First we find the midpoint of the diameter: $(h, k) = \left(\dfrac{-3 + 5}{2}, \dfrac{2 + 6}{2} \right) = (1, 4)$. These are the coordinates of the center. Next use the center and one endpoint of the diameter in the distance formula to find the radius: $r = \sqrt{(5 - 1)^2 + (6 - 4)^2} = \sqrt{16 + 4} = \sqrt{20}$. Putting these values into the standard form for a circle, we get $(x - 1)^2 + (y - 4)^2 = 20$.

26. Find the equation of the circle whose center is $(4, -2)$ and which passes through the point $(-4, 13)$.

We have $h = 4$ and $k = -2$. Using the point on the circle, we can calculate the radius: $r = \sqrt{(-4 - 4)^2 + (13 - (-2))^2} = \sqrt{64 + 225} = \sqrt{289} = 17$. Next put these values into the standard form of a circle: $(x - 4)^2 + (y + 2)^2 = 289$.

28. Find the equation of the circle whose center lies on the y axis and which passes through the two points $(2, 2)$ and $(6, -4)$.

We know that $h = 0$ because the center is on the y axis. Call the y coordinate k. Then the distance from the center to either of the points on the circle must be the same:

$$\begin{aligned}
\sqrt{(0-2)^2 + (k-2)^2} &= \sqrt{(0-6)^2 + (k-(-4))^2} \\
\sqrt{4 + (k-2)^2} &= \sqrt{36 + (k+4)^2} \\
4 + (k-2)^2 &= 36 + (k+4)^2 \\
4 + k^2 - 4k + 4 &= 36 + k^2 + 8k + 16 \\
k^2 - 4k + 8 &= k^2 + 8k + 52 \\
8k + 52 &= -4k + 8 \\
8k &= -4k - 44 \\
12k &= -44 \\
k &= -\frac{11}{3}
\end{aligned}$$

Now we need the radius: $r = \sqrt{(0-2)^2 + \left(-\dfrac{11}{3} - 2\right)^2} = \sqrt{4 + \dfrac{289}{9}} = \sqrt{\dfrac{325}{9}}$. Therefore the equation of the circle is $x^2 + \left(y + \dfrac{11}{3}\right)^2 = \dfrac{325}{9}$.

30. The equation of a circle is $(x+2)^2 + (y-3)^2 = 25$. Find the equation of the line passing through the point $(3, 3)$ and tangent to this circle. From problem 29, we know that the equation of the tangent line to the circle $x^2 + y^2 = r^2$ at the point $P_1(x_1, y_1)$ on the circle is given by $x_1 x + y_1 y = r^2$. If we replace x with $x - h$ and y with $y - k$ then we get the equation $x_1(x - h) + y_1(y - k) = r^2$. For this circle we have $h = -2$, $k = 3$, $r = 5$, $x_1 = 3$, and $y_1 = 3$. Therefore the equation of the tangent line is

$$\begin{aligned}
3(x+2) + 3(y-3) &= 25 \\
3x + 6 + 3y - 9 &= 25 \\
3x + 3y - 3 &= 25 \\
3x + 3y &= 28
\end{aligned}$$

8.3   Exercises

In exercises 1 - 10, use the method of completing the square to put each parabola into standard form. Then identify the coordinates of the vertex and graph the resulting parabola.

The standard form of a parabola that opens up or down with vertex located at (h, k) is $y = a(x-h)^2 + k$. If the parabola opens left or right then the equation becomes $x = a(y-k)^2 + h$.

2. $y = x^2 - 4x + 6$

First subtract 6 from both sides of the equation:

$y - 6 = x^2 - 4x$

Now take half of -4 and square it, giving 4. Add 4 to both sides:

$y - 6 + 4 = x^2 - 4x + 4$

Simplify the left hand side and factor the right hand side:

$y - 2 = (x - 2)^2$ Solve for y:

$y = (x - 2)^2 + 2$ The coordinates of the vertex are $(2, 2)$ and the graph opens up since $a > 0$.

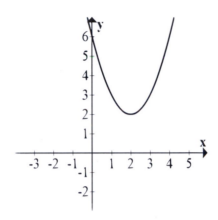

4. $y = 2x^2 + 6x - 3$

First add 3 to both sides:

$y + 3 = 2x^2 + 6x$

Next factor 2 from the right hand side:

$y + 3 = 2(x^2 + 3x)$

Now square half of 3. Add this inside the parentheses, but add twice this to the left hand side:

$y + 3 + \dfrac{9}{2} = 2\left(x^2 + 3x + \dfrac{9}{4}\right)$

Simplify the left hand side and factor the right hand side:

$y + \dfrac{15}{2} = 2\left(x + \dfrac{3}{2}\right)^2$

Solve for y:

$$y = 2\left(x + \frac{3}{2}\right)^2 - \frac{15}{2}$$

The vertex is located at $\left(-\frac{3}{2}, -\frac{15}{2}\right)$ and the graph opens up:

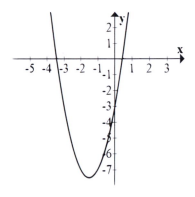

8. $x = 2y^2 + 8y - 6$

First add 6 to both sides:

$x + 6 = 2y^2 + 8y$

Now factor 2 out of the right hand side:

$x + 6 = 2(y^2 + 4y)$

Next square half of 4, getting 4. Add 4 inside the parentheses and twice this to the left hand side:

$x + 6 + 8 = 2(y^2 + 4y + 4)$

Simplify the left hand side and factor the right:

$x + 14 = 2(y + 2)^2$

Solve for x:

$x = 2(y + 2)^2 - 14$ The vertex is located at $(14, -2)$ and the parabola opens to the right:

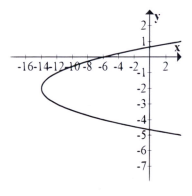

In problems 11 - 24, find the equation of the parabola satisfying the given conditions.

If we are given the distance from the vertex to the focus, or the vertex to the directrix, we call this variable p. Then in the standard form for the parabola, we use the formula $a = \dfrac{1}{4p}$.

12. vertex at the origin; focus at $\left(0, \dfrac{1}{4}\right)$

Since the focus is above the vertex, the parabola opens up. Furthermore, $p = \dfrac{1}{4}$. Therefore $a = 1$. We have $h = k = 0$, so the equation of the parabola is $y = x^2$.

14. vertex at the origin; focus at $(2, 0)$

Since the focus is to the right of the vertex, the parabola opens to the right. We have $p = 2$ so $a = \dfrac{1}{4(2)} = \dfrac{1}{8}$. Therefore the equation of the parabola is $x = \dfrac{y^2}{8}$, or $y^2 = 8x$.

16. vertex at $(2, 3)$; focus at $(2, 2)$

Since the focus is below the vertex, the parabola opens down. We have $p = 1$, $h = 2$, and $k = 3$. Therefore $a = -\dfrac{1}{4}$ and the equation of the parabola is $y = -\dfrac{1}{4}(x - 2)^2 + 3$.

18. vertex at the origin; passing through the points $(-1, 2)$ and $(1, 2)$

Since the vertex is at the origin, we have $h = k = 0$. The two points are to the left and right of the vertex, and above it. Therefore the parabola opens up and must be of the form $y = a(x - h)^2 + k$. Using $h = k = 0$ and $x = 1$, $y = 2$, we can substitute these values into the equation. This yields $2 = a(1 - 0)^2 + 0$, or $a = 2$. Therefore the equation of the parabola is $y = 2x^2$.

20. vertex at $(2, -2)$; passing through the points $(3, -1)$ and $(6, 0)$

A graph of these points appears below:

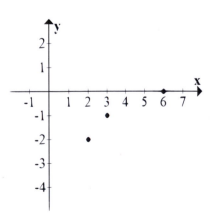

Since $(2, -2)$ is the vertex, it seems like the parabola opens to the right. We have $h = 2$ and $k = -2$. We need a. The equation is of the form $x = a(y - k)^2 + h$. Substituting h and k along with the point $x = 3$, $y = -1$, we get $3 = a(-1 - (-2))^2 + 2$. This becomes $3 = a + 2$, so $a = 1$. Therefore the equation of the parabola is $x = (y + 2)^2 + 2$. Checking both points in the equation verifies its accuracy.

22. vertex at $(5, 4)$; directrix $y = 0$

Since the directrix is below the vertex, the parabola opens up. The distance from the vertex to the directrix is $4 - 0 = 4$. Therefore $p = 4$ and $a = \dfrac{1}{4(4)} = \dfrac{1}{16}$. We have $h = 5$ and $k = 4$. The equation of the parabola is $y = \dfrac{1}{16}(x - 5)^2 + 4$.

24. focus at $(5, 4)$; directrix $x = 1$

Since the focus is to the right of the directrix, the parabola opens to the right. The distance from the focus to the directrix is $5 - 1 = 4$. Therefore the distance from the vertex to the focus is half this, and $p = 2$. This gives $a = \dfrac{1}{4(2)} = \dfrac{1}{8}$. The standard form of the equation of a parabola opening to the right is $x = a(y - k)^2 + h$, so the equation of this parabola is $x = \dfrac{1}{8}(y - 4)^2 + 5$.

26. Find the equation of the parabola passing through the points $(-1, 6)$, $(1, 0)$, and $(2, 3)$.

For this parabola we do not have any information about the vertex, focus, or directrix. We will use the general form for the equation of a parabola, and substitute the three points into this equation. If we graph the three points it appears that the parabola opens up, so we will use the equation $y = ax^2 + bx + c$:

$$\begin{cases} 6 & = & a(-1)^2 + b(-1) + c \\ 0 & = & a(1^2) + b(1) + c \\ 3 & = & a(2^2) + b(2) + c \end{cases}$$

Simplifying the equations:

$$\begin{cases} 6 & = & a - b + c \\ 0 & = & a + b + c \\ 3 & = & 4a + 2b + c \end{cases}$$

If we subtract the second equation from the first equation, we get $-2b = 6$, so $b = -3$. Substituting this into the first and third equations and simplifying gives

$$\begin{cases} 3 & = & a + c \\ 9 & = & 4a + c \end{cases}$$

Subtracting the first equation from the second, we get $3a = 6$, so $a = 2$. Therefore $c = 1$ and the equation of the parabola is $y = 2x^2 - 3x + 1$.

28. Find the equation of the tangent line to $y^2 + 4x + 2y + 9 = 0$ at the point $(-6, 3)$.

The equation of the line tangent to the parabola $y^2 = 4px$ at any point $P_1(x_1, y_1)$ on the parabola can be written as $y_1 y = 2p(x + x_1)$. Substituting $y - k$ in for y and $x - h$ in for x, we get $y_1(y - k) = 2p((x - h) + x_1)$. Our parabola is $y^2 + 4x + 2y + 9 = 0$. Subtracting both $4x$ and 9 from both sides of the equation gives $y^2 + 2y = -4x - 9$. Adding 1 to both sides we get $y^2 + 2y + 1 = -4x - 8$. Factoring the left hand side and adding 8, we get $(y + 1)^2 + 8 = -4x$. Dividing by -4, we get $x = -\dfrac{1}{4}(y + 1)^2 - 2$. Therefore $h = -2$ and $k = -1$. Since $a = \dfrac{1}{4}$, we have $p = -1$. Finally, from the point that is given in the problem, we have $x_1 = -6$ and $y_1 = 3$. Therefore the equation of the tangent line is

$$
\begin{aligned}
y_1(y - k) &= 2p((x - h) + x_1) \\
3(y - (-1)) &= 2(-1)((x - (-2)) + (-6)) \\
3(y + 1) &= -2(x + 2 - 6) \\
3y + 3 &= -2x + 8 \\
2x + 3y &= 5
\end{aligned}
$$

8.4 Exercises

In exercises 1 - 10, place each equation in standard form using completing the square. Then identify the coordinates of the center, the lengths of the major and minor axes, and draw a graph of the resulting ellipse.

The standard form for the equation of a horizontal ellipse with major axis length $2a$ and minor axis length $2b$, centered at the point (h, k), is

$$
\frac{(x - h)^2}{a^2} + \frac{(y - k)^2}{b^2} = 1
$$

If it is a vertical ellipse then the equation becomes

$$
\frac{(x - h)^2}{b^2} + \frac{(y - k)^2}{a^2} = 1
$$

2. $9x^2 + 4y^2 = 36$

Dividing both sides by 36, we get $\dfrac{x^2}{4} + \dfrac{y^2}{9} = 1$. Therefore $a = 3$ and $b = 2$. The center of the ellipse is the origin:

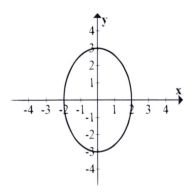

6. $25x^2 + 4y^2 - 100x + 16y + 16 = 0$

First we group the x and y terms together, and subtract 16 from both sides:

$(25x^2 - 100x) + (4y^2 + 16y) = -16$

Next factor 25 from the first parentheses, and 4 from the second parentheses:

$25(x^2 - 4x) + 4(y^2 + 4y) = -16$.

We need to add 4 inside the first parentheses in order to completet the square. Since there is a coefficient of 25 outside the parentheses, we will actually add 100 to the right hand side:
$25(x^2 - 4x + 4) + 4(y^2 + 4y) = -16 + 100$

Now we add 4 inside the second parentheses and 16 to the right hand side:

$25(x^2 - 4x + 4) + 4(y^2 + 4y + 4) = 100$

Next factor both sets of parentheses:

$25(x - 2)^2 + 4(y + 2)^2 = 100$

Finally, divide both sides by 100:

$\dfrac{(x - 2)^2}{4} + \dfrac{(y + 2)^2}{25} = 1$ The center of the ellipse is $(2, -2)$. We have $a = 5$ and $b = 2$, and it is a vertical ellipse:

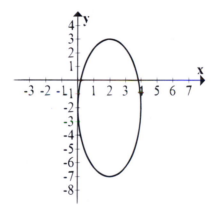

In exercises 11 - 20, find the equation of the ellipse satisfying the given conditions.

The semimajor axis of an ellipse has length a, the semiminor axis has length b, the distance from the center to either focus is c, and $a^2 = b^2 + c^2$. Remember that the foci of an ellipse always lie on the major axis.

12. center at the origin; foci at $(\pm 3, 0)$; semimajor axis length 5

Since the center is at the origin and one focus is at the point $(3, 0)$, we know that $c = 3$ and we have a horizontal ellipse. Furthermore, $a = 5$. Therefore since $a^2 = b^2 + c^2$, we can calculate $b = 4$. We have $h = k = 0$. The equation of the ellipse is $\dfrac{x^2}{25} + \dfrac{y^2}{16} = 1$.

14. center at the origin; foci at $(0, \pm 6)$; semiminor axis length 8

The foci are above and below the center, so we have a vertical ellipse. Furthermore, $c = 6$ and $b = 8$. Therefore $a^2 = 8^2 + 6^2$ so $a = 10$. The equation of the ellipse is $\dfrac{x^2}{64} + \dfrac{y^2}{100} = 1$.

16. center at the origin; passing through the points $(\pm 4, 0)$ and $(0, \pm 5)$

Since the center is at the origin, the first points given are the endpoints of the minor axis and the second points given are the endpoints of the major axis. Therefore $b = 4$ and $a = 5$. The ellipse is vertical, so the equation of the ellipse is $\dfrac{x^2}{16} + \dfrac{y^2}{25} = 1$.

18. center at the point $(-2, 3)$; one focus at $(-2, 7)$; semiminor axis length 3

The given focus is above the center, so the ellipse is vertical. We have $c = 7 - 3 = 4$. The semiminor axis has length 3, so $b = 3$. Therefore $a = 5$ and the equation of the ellipse is $\dfrac{(x + 2)^2}{9} + \dfrac{(y - 3)^2}{25} = 1$.

20. endpoints of the minor axis located at $(0, 2)$ and $(6, 2)$; focus at $(3, 6)$

The midpoint of the minor axis is $\left(\dfrac{0 + 6}{2}, \dfrac{2 + 2}{2} \right) = (3, 2)$, so $h = 3$ and $k = 2$. Furthermore the semiminor axis has length $b = \dfrac{6 - 0}{2} = 3$. The focus is 4 units above the center, so $c = 4$. Finally, $a^2 = 3^2 + 4^2$, so $a = 5$. This is a vertical ellipse. The equation of the ellipse is $\dfrac{(x - 3)^2}{9} + \dfrac{(y - 2)^2}{25} = 1$.

24. Find the equation of the tangent line to the ellipse $\dfrac{(x - 2)^2}{25} + \dfrac{(y + 1)^2}{16} = 1$ at the point $\left(6, \dfrac{7}{5} \right)$.

The equation of the tangent line to the ellipse $\dfrac{x^2}{a^2} + \dfrac{y^2}{b^2} = 1$ at the point $P_1(x_1, y_1)$ on the ellipse can be written as $b^2 x_1 x + a^2 y_1 y = a^2 b^2$. Substituting $(x - h)$ for x and $(y - k)$ for y, we get the equation $b^2 x_1(x - h) + a^2 y_1(y - k) = a^2 b^2$. For the given ellipse we have $h = 2$, $k = -1$, $a = 5$, and $b = 4$. From the point given, we have $x_1 = 6$ and $y_1 = \dfrac{7}{5}$.

Therefore the equation of the tangent line is

$$
\begin{aligned}
b^2 x_1(x - h) + a^2 y_1(y - k) &= a^2 b^2 \\
16(6)(x - 2) + 25\left(\frac{7}{5}\right)(y - (-1)) &= (25)(16) \\
96(x - 2) + 35(y + 1) &= 400 \\
96x - 192 + 35y + 35 &= 400 \\
96x + 35y &= 627
\end{aligned}
$$

8.5   Exercises

In problems 1 - 10, put each hyperbola into standard form. Identify the coordinates of the center and graph.

The standard form for the equation of a hyperbola centered at (h, k) that opens left and right is

$$\frac{(x - h)^2}{a^2} - \frac{(y - k)^2}{b^2} = 1$$

If it opens up and down then the equation becomes

$$\frac{(y - k)^2}{a^2} - \frac{(x - h)^2}{b^2} = 1$$

c is still the distance from the center to one of the foci, but this time $c^2 = a^2 + b^2$. a is the distance from the center to one of the vertices of the hyperbola.

2. $9x^2 - 4y^2 = 36$

We divide both sides by 36 and obtain $\dfrac{x^2}{4} - \dfrac{y^2}{9} = 1$. The center is located at the origin. A graph appears below:

6. $25x^2 - 4y^2 - 100x + 16y - 16 = 0$

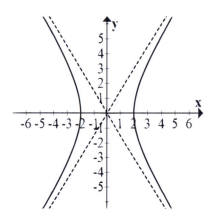

We start by grouping the x terms and the y terms, then add 16 to both sides:

$(25x^2 - 100x) - (4y^2 - 16y) = 16$

Next we factor 25 out of the first parentheses, and 4 out of the second parentheses:

$25(x^2 - 4x) - 4(y^2 - 4y) = 16$

We need to add 4 into the first set of parentheses. This number gets multiplied by 25, so we add 100 to the right hand side:

$25(x^2 - 4x + 4) - 4(y^2 - 4y) = 16 + 100$

When we add 4 into the second set of parentheses, it gets multiplied by -4, so we subtract 16 from the right hand side:

$25(x^2 - 4x + 4) - 4(y^2 - 4y + 4) = 100$

Now factor both sets of parentheses:

$25(x - 2)^2 - 4(y - 2)^2 = 100$

Finally, divide by 100:

$$\frac{(x - 2)^2}{4} - \frac{(y - 2)^2}{25} = 1$$

The center of the hyperbola is located at the point $(2, 2)$ and it opens left and right:

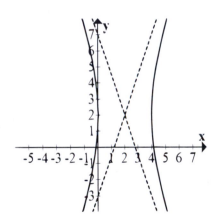

In exercises 11 - 20, find the equation of the hyperbola satisfying the given conditions.

12. center at the origin; foci at $(\pm 13, 0)$; vertices at $(\pm 5, 0)$

Since the center is at the origin and one focus is at $(13, 0)$, we know that $c = 13$. Furthermore, on vertex is at $(5, 0)$, so $a = 5$. This gives $13^2 = 5^2 + b^2$, so $b = 12$. We also have $h = k = 0$. The equation of the hyperbola in standard form is $\dfrac{x^2}{25} - \dfrac{y^2}{144} = 1$.

14. center at the origin; foci at $(0, \pm 17)$; vertices at $(0, \pm 15)$

We have $c = 17$ and $a = 15$. This gives $17^2 = 15^2 + b^2$, so $b = 8$. The hyperbola opens up and down (the vertices are above and below the center), so the variable y shows up first in the equation, which is $\dfrac{y^2}{225} - \dfrac{x^2}{64} = 1$.

16. center at the origin; passing through the points $(\pm 4, 0)$ and $\left(\pm\dfrac{20}{3}, \pm 4\right)$

Since the center is at the origin, we know that $h = k = 0$. The vertices are at $(\pm 4, 0)$, so it is a horizontal hyperbola and $a = 4$. Putting these into the standard form for a hyperbola, we get the equation $\dfrac{x^2}{16} - \dfrac{y^2}{b^2} = 1$. To solve for b we can substitute the point $\left(\dfrac{20}{3}, 4\right)$ into the equation. This gives

$$
\begin{aligned}
\frac{\left(\frac{20}{3}\right)^2}{16} - \frac{3^2}{b^2} &= 1 \\
\frac{400}{144} - \frac{9}{b^2} &= 1 \\
-\frac{9}{b^2} &= -\frac{256}{400} \\
256b^2 &= 3600 \\
b^2 &= \frac{225}{16} \\
b &= \frac{5}{4}
\end{aligned}
$$

Therefore the equation of the hyperbola is

$$
\begin{aligned}
\frac{x^2}{16} - \frac{y^2}{\left(\frac{5}{4}\right)^2} &= 1 \\
\frac{x^2}{16} - \frac{y^2}{\frac{25}{16}} &= 1 \\
\frac{x^2}{16} - \frac{16y^2}{25} &= 1
\end{aligned}
$$

18. center at the point $(-2, 3)$; foci at $(-2, 13)$ and $(-2, -7)$; vertices at $(-2, 9)$ and $(-2, -3)$

We know immediately that $h = -2$ and $k = 3$. The foci are above and below the center, so the hyperbola opens up and down. We have $c = 13 - 3 = 10$. The distance from the center to one of the vertices is $a = 9 - 3 = 6$. Now we use the equation $c^2 = a^2 + b^2$ to get $b = 8$. Therefore the equation of the hyperbola is $\dfrac{(y-3)^2}{36} - \dfrac{(x+2)^2}{64} = 1$.

20. foci at $(0, \pm 13)$; asymptotes at $y = \pm\dfrac{12}{5}x$

The point halfway between the two foci is the origin. Therefore $h = k = 0$ and $c = 13$. The equations of the asymptotes are $y = \pm\dfrac{ax}{b}$. Therefore $\dfrac{a}{b} = \dfrac{12}{5}$, and $a^2 + b^2 = 13$, so $a = 12$ and $b = 5$. The equation of the hyperbola is $\dfrac{y^2}{144} - \dfrac{x^2}{25} = 1$.

8.6   Exercises

In each of the following, identify the conic section. Then put the equation into standard form. Write the appropriate translation of axes which will locate the center of the conic section at the new origin, and graph the resulting conic.

If we have the equation of a conic section in general form, that is, $Ax^2 + By^2 + Cx + Dy + E = 0$, then it is a circle if $A = B$, a parabola if $AB = 0$, an ellipse if $AB > 0$ (and $A \neq B$), and a hyperbola if $AB < 0$.

2. $x^2 + y^2 - 6x - 4y - 12 = 0$

This is a circle because $A = B$. We first add 12 to both sides, and group the variables together:

$(x^2 - 6x) + (y^2 - 4y) = 12$

Now we add 9 inside the first parentheses, and 4 inside the second. Add the same numbers to the right hand side:

$(x^2 - 6x + 9) + (y^2 - 4y + 4) = 12 + 9 + 4$

Now factor the left hand side, and combine the right hand side:

$(x - 3)^2 + (y - 2)^2 = 25$

The center of the circle is located at the point $(3, 2)$, and the radius is 5:

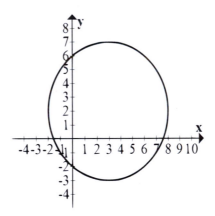

8. $x^2 - 4x - y + 3 = 0$

We have $A = 1$ and $B = 0$, so this is a parabola. First move the y to the other side of the equation:

$y = x^2 - 4x + 3$

Next subtract 3 from both sides

$y - 3 = x^2 - 4x$

Now take half of -4 and square it, getting 4. So we add 4 to both sides:

$y - 3 + 4 = x^2 - 4x + 4$

Simplify the left hand side and factor the right:

$y + 1 = (x - 2)^2$

Solve for y:

$y = (x - 2)^2 - 1$ The vertex is located at the point $(2, -1)$, and the parabola opens up:

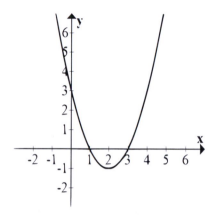

14. $25x^2 + 4y^2 - 100x + 16y + 16 = 0$

This is an ellipse because $A \neq B$ and they are both positive. First subtract 16 from both sides:

$25x^2 + 4y^2 - 100x + 16y = -16$

Now group the x terms and the y terms:

$(25x^2 - 100x) + (4y^2 + 16y) = -16$

Factor 25 out of the first parentheses and 4 out of the second pair:

$25(x^2 - 4x) + 4(y^2 + 4y) = -16$

Add 4 inside the both pairs of parentheses. THen we add 100 and 16 to the right hand side:

$25(x^2 - 4x + 4) + 4(y^2 + 4y + 4) = -16 + 100 + 16$

Simplify the right hand side, and factor the left hand side:

$25(x - 2)^2 + 4(y + 2)^2 = 100$

Divide by 100:

$\dfrac{(x - 2)^2}{4} + \dfrac{(y + 2)^2}{25} = 1$

The ellipse is a vertical ellipse, and the center is at the point $(2, -2)$.

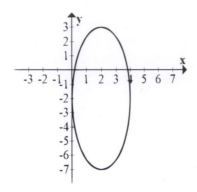

20. $25x^2 - 4y^2 - 100x + 16y - 16 = 0$

Since A and B have opposite signs, this is a hyperbola. First we add 16 to both sides:

$25x^2 - 4y^2 - 100x + 16y = 16$

Now group the x terms together, as well as the y terms:

$(25x^2 - 100x) - (4y^2 - 16y) = 16$

Next factor the leading coefficient from each set of parentheses:

$25(x^2 - 4x) - 4(y^2 - 4y) = 16$

We need to add 4 inside each pair of parentheses. The first pair is multiplied by 25, so we add 100 to the right hand side. The second pair is multiplied by -4, so we subtract 16:

$25(x^2 - 4x + 4) - 4(y^2 - 4y + 4) = 16 + 100 - 16$

Now factor the left hand side and simplify the right:

$25(x - 2)^2 - 4(y - 2)^2 = 100$

Finally, divide both sides by 100:

$$\frac{(x - 2)^2}{4} - \frac{(y - 2)^2}{25} = 1$$

The center is located at the point $(2, 2)$, and the hyperbola opens left and right:

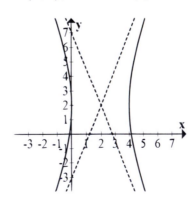

8.7   Exercises

In problems 1 - 10, use the formula $\cot 2\theta = \dfrac{A - C}{B}$ *to calculate* θ. *Then substitute* θ *into the formulas* $x' = x\cos\theta + y\sin\theta$ *and* $y' = -x\sin\theta + y\cos\theta$.

For this formula, remember that A is the coefficient of x^2, B is the coefficient of xy, and C is the coefficient of y^2.

2. $x^2 - 4xy + y^2 + 3 = 0$

We have $A = 1$, $B = -4$, and $C = 1$. Therefore $\cot 2\theta = \dfrac{A - C}{B} = \dfrac{1 - 1}{-4} = 0$ and $2\theta = \dfrac{\pi}{2}$, so $\theta = \dfrac{\pi}{4}$. This gives

$$
\begin{aligned}
x' &= x\cos\theta + y\sin\theta \\
&= x\cos\frac{\pi}{4} + y\sin\frac{\pi}{4} \\
&= \frac{\sqrt{2}}{2}(x + y) \\
y' &= -x\sin\theta + y\cos\theta \\
&= -x\sin\frac{\pi}{4} + y\cos\frac{\pi}{4} \\
&= \frac{\sqrt{2}}{2}(y - x)
\end{aligned}
$$

8. $25x^2 - 36xy + 40y^2 - 12\sqrt{13}x - 8\sqrt{13}y = 0$

We have $A = 25$, $B = -36$, and $C = 40$. Therefore $\cot 2\theta = \dfrac{A - C}{B} = \dfrac{25 - 40}{-36} = \dfrac{5}{12}$.
Since $\cot 2\theta = \dfrac{\cos 2\theta}{\sin 2\theta}$, we can conclude that $\cos 2\theta = \dfrac{5}{13}$ and $\sin 2\theta = \dfrac{12}{13}$. Therefore
$\cos\theta = \sqrt{\dfrac{1 + \cos 2\theta}{2}} = \sqrt{\dfrac{1 + \frac{5}{13}}{2}} = \sqrt{\dfrac{9}{13}} = \frac{3\sqrt{13}}{13}$. Similarly, $\sin\theta = \dfrac{2\sqrt{13}}{13}$. Therefore

$$
\begin{aligned}
x' &= x\cos\theta + y\sin\theta \\
&= x\left(\frac{3\sqrt{13}}{13}\right) + y\left(\frac{2\sqrt{13}}{13}\right) \\
y' &= -x\sin\theta + y\cos\theta \\
&= -x\left(\frac{2\sqrt{13}}{13}\right) + y\left(\frac{3\sqrt{13}}{13}\right)
\end{aligned}
$$

10. $34x^2 - 24xy + 41y^2 - 25 = 0$

We have $A = 34$, $B = -24$, and $C = 41$. Therefore $\cot 2\theta = \dfrac{A - C}{B} = \dfrac{34 - 41}{-24} = \dfrac{7}{24}$.
Since $\cot 2\theta = \dfrac{\cos 2\theta}{\sin 2\theta}$, we can conclude that $\cos 2\theta = \dfrac{7}{25}$ and $\sin 2\theta = \dfrac{24}{25}$. This leads to

$$\cos\theta = \sqrt{\frac{1+\cos 2\theta}{2}} = \sqrt{\frac{1+\frac{7}{25}}{2}} = \sqrt{\frac{16}{25}} = \frac{4}{5}.$$ Therefore $\sin\theta = \frac{3}{5}$. We have

$$
\begin{aligned}
x' &= x\cos\theta + y\sin\theta \\
&= x\left(\frac{4}{5}\right) + y\left(\frac{3}{5}\right) \\
&= \frac{4x + 3y}{5} \\
y' &= -x\sin\theta + y\cos\theta \\
&= -x\left(\frac{3}{5}\right) + y\left(\frac{4}{5}\right) \\
&= \frac{4y - 3x}{5}
\end{aligned}
$$

In problems 11 - 20, calculate the discriminant of the quadratic form using the formula $4AC - B^2$. *Then decide what type of conic each equation represents, based on the value of the discriminant (note: these are the same equations from problems 1 - 10).*

The discriminant of a quadratic form determines the type of conic the graph will take. In particular, if $4AC - B^2 < 0$ then the conic is a hyperbola. If $4AC - B^2 = 0$ then the conic is a parabola, and if $4AC - B^2 > 0$ then the conic is an ellipse.

12. $x^2 - 4xy + y^2 + 3 = 0$

We have $4AC - B^2 = 4(1)(1) - (-4)^2 = -12$, so this conic is a hyperbola.

18. $25x^2 - 36xy + 40y^2 - 12\sqrt{13}x - 8\sqrt{13}y = 0$

We have $4AC - B^2 = 4(25)(40) - (-36)^2 = 2704$, so this conic is an ellipse.

20. $16x^2 + 24xy + 9y^2 - 60x + 80y = 0$

We have $4AC - B^2 = 4(16)(9) - 24^2 = 0$, so this conic is a parabola.

In problems 21 - 30, calculate the values of A', C', D', E' *and* F'. *Draw the standard axes* (x, y) *and the rotated axes* (x', y'), *then graph the resulting conic on the rotated set of axes (note: these are the same equations from problems 1 - 10).*

We will use the following formulas:

$$
\begin{aligned}
A' &= A\cos^2\theta + B\cos\theta\sin\theta + C\sin^2\theta \\
C' &= A\sin^2\theta - B\sin\theta\cos\theta + C\cos^2\theta \\
D' &= D\cos\theta + E\sin\theta \\
E' &= -D\sin\theta + E\cos\theta \\
F' &= F
\end{aligned}
$$

22. $x^2 - 4xy + y^2 + 3 = 0$

We have $A = 1$, $B = -4$, $C = 1$, $D = 0$, $E = 0$, $F = 3$, $\cos\theta = \cos\dfrac{\pi}{4} = \dfrac{\sqrt{2}}{2}$ and

$\sin\theta = \sin\dfrac{\pi}{4} = \dfrac{\sqrt{2}}{2}$. Therefore

$$
\begin{aligned}
A' &= A\cos^2\theta + B\cos\theta\sin\theta + C\sin^2\theta \\
&= (1)\left(\frac{\sqrt{2}}{2}\right)^2 + (-4)\left(\frac{\sqrt{2}}{2}\right)\left(\frac{\sqrt{2}}{2}\right) + (1)\left(\frac{\sqrt{2}}{2}\right)^2 \\
&= \frac{1}{2} - \frac{4}{2} + \frac{1}{2} \\
&= -1 \\
C' &= A\sin^2\theta - B\sin\theta\cos\theta + C\cos^2\theta \\
&= (1)\left(\frac{\sqrt{2}}{2}\right)^2 - (-4)\left(\frac{\sqrt{2}}{2}\right)\left(\frac{\sqrt{2}}{2}\right) + (1)\left(\frac{\sqrt{2}}{2}\right)^2 \\
&= \frac{1}{2} + 4\left(\frac{1}{2}\right) + \frac{1}{2} \\
&= 3 \\
D' &= D\cos\theta + E\sin\theta \\
&= (0)\left(\frac{\sqrt{2}}{2}\right) + (0)\left(\frac{\sqrt{2}}{2}\right) \\
&= 0 \\
E' &= -D\sin\theta + E\cos\theta \\
&= -(0)\left(\frac{\sqrt{2}}{2}\right) + (0)\left(\frac{\sqrt{2}}{2}\right) \\
&= 0 \\
F' &= 3
\end{aligned}
$$

The new equation is $A'(x')^2 + C'(y')^2 + D'x' + E'y' + F' = -(x')^2 + 3(y')^2 + 3 = 0$. Subtracting 3 from both sides, and dividing by -3, we get the equation $\dfrac{(x')^2}{3} - (y')^2 = 1$. A graph appears on the next page:

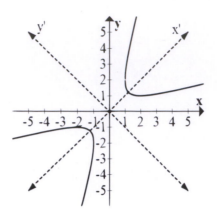

28. $25x^2 - 36xy + 40y^2 - 12\sqrt{13}x - 8\sqrt{13}y = 0$

We have $A = 25$, $B = -36$, $C = 40$, $D = -12\sqrt{13}$, $E = -8\sqrt{13}$, $F = 0$, $\cos\theta = \dfrac{3\sqrt{13}}{13}$, and

$\sin\theta = \dfrac{2\sqrt{13}}{13}$. Therefore

$$
\begin{aligned}
A' &= A\cos^2\theta + B\cos\theta\sin\theta + C\sin^2\theta \\
&= 25\left(\frac{3\sqrt{13}}{13}\right)^2 - 36\left(\frac{3\sqrt{13}}{13}\right)\left(\frac{2\sqrt{13}}{13}\right) + 40\left(\frac{2\sqrt{13}}{13}\right)^2 \\
&= \frac{25(9)}{13} - \frac{36(3)(2)}{13} + \frac{40(4)}{13} \\
&= 13 \\
C' &= A\sin^2\theta - B\sin\theta\cos\theta + C\cos^2\theta \\
&= 25\left(\frac{2\sqrt{13}}{13}\right)^2 - (-36)\left(\frac{2\sqrt{13}}{13}\right)\left(\frac{3\sqrt{13}}{13}\right) + 40\left(\frac{3\sqrt{13}}{13}\right)^2 \\
&= \frac{25(4)}{13} + \frac{36(2)(3)}{13} + \frac{40(9)}{13} \\
&= 52 \\
D' &= D\cos\theta + E\sin\theta \\
&= (-12\sqrt{13})\left(\frac{3\sqrt{13}}{13}\right) + (-8\sqrt{13})\left(\frac{2\sqrt{13}}{13}\right) \\
&= (-12)(3) + (-8)(2) \\
&= -52 \\
E' &= -D\sin\theta + E\cos\theta \\
&= -(-12\sqrt{13})\left(\frac{2\sqrt{13}}{13}\right) + (-8\sqrt{13})\left(\frac{3\sqrt{13}}{13}\right) \\
&= 0 \\
F' &= 0
\end{aligned}
$$

The new equation is $A'(x')^2 + C'(y')^2 + D'x' + E'y' + F' = 13(x')^2 + 52(y')^2 - 4x'\sqrt{2} = 0$. In order to put this ellipse into standard form, we first divide by 13, then group like variables and complete the square:

$$
\begin{aligned}
(x')^2 + 4(y')^2 - 4x' &= 0 \\
\left((x')^2 - 4x'\right) + 4(y')^2 &= 0 \\
\left((x')^2 - 4x + 4\right) + 4(y')^2 &= 4 \\
\frac{(x'-2)^2}{4} + (y')^2 &= 1
\end{aligned}
$$

A graph appears below:

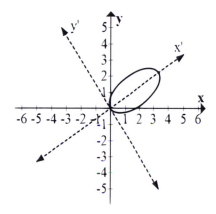

30. $16x^2 + 24xy + 9y^2 - 60x + 80y = 0$

We have $A = 16$, $B = 24$, $C = 9$, $D = -60$, $E = 80$, $F = 0$, $\cos\theta = \dfrac{4}{5}$, and $\sin\theta = \dfrac{3}{5}$. Therefore

$$
\begin{aligned}
A' &= A\cos^2\theta + B\cos\theta\sin\theta + C\sin^2\theta \\
&= 16\left(\frac{4}{5}\right)^2 + (24)\left(\frac{4}{5}\right)\left(\frac{3}{5}\right) + 9\left(\frac{3}{5}\right)^2 \\
&= \frac{16(4) + 24(4)(3) + 9(9)}{25} \\
&= \frac{433}{25} \\
C' &= A\sin^2\theta - B\sin\theta\cos\theta + C\cos^2\theta \\
&= 16\left(\frac{3}{5}\right)^2 - (24)\left(\frac{3}{5}\right)\left(\frac{4}{5}\right) + 9\left(\frac{4}{5}\right)^2 \\
&= \frac{16(9) - 24(3)(4) + 9(16)}{25} \\
&= 0
\end{aligned}
$$

$$
\begin{aligned}
D' &= D\cos\theta + E\sin\theta \\
&= (-60)\left(\frac{4}{5}\right) + (80)\left(\frac{3}{5}\right) \\
&= \frac{(-60)(4) + (80)(3)}{5} \\
&= 0 \\
E' &= -D\sin\theta + E\cos\theta \\
&= -(-60)\left(\frac{3}{5}\right) + (80)\left(\frac{4}{5}\right) \\
&= \frac{60(3) + 80(4)}{5} \\
&= 100 \\
F' &= 0
\end{aligned}
$$

The new equation is $A'(x')^2 + C'(y')^2 + D'x' + E'y' + F' = \dfrac{433}{25}(x')^2 + 100y' = 0$. Solving for y', we get $y' = -\dfrac{433(x')^2}{2500}$. A graph appears below:

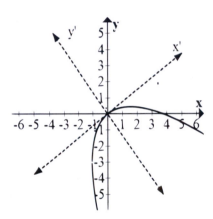

Chapter 9

Sequences and Series

9.1   Exercises

In problems 1-10, write the first five terms of the sequence (Assume n begins with 1). For these problems we treat a_n like a function, and n is the independent variable.

2. $a_n = 4n - 3$

   $a_1 = 4(1) - 3 = 1$

   $a_2 = 4(2) - 3 = 5$

   $a_3 = 4(3) - 3 = 9$

   $a_4 = 4(4) - 3 = 13$

   $a_5 = 4(5) - 3 = 17$

4. $a_n = \left(\dfrac{1}{3}\right)^n$

   $a_1 = \left(\dfrac{1}{3}\right)^1 = \dfrac{1}{3}$

   $a_2 = \left(\dfrac{1}{3}\right)^2 = \dfrac{1}{9}$

   $a_3 = \left(\dfrac{1}{3}\right)^3 = \dfrac{1}{27}$

   $a_4 = \left(\dfrac{1}{3}\right)^4 = \dfrac{1}{81}$

   $a_5 = \left(\dfrac{1}{3}\right)^5 = \dfrac{1}{243}$

6. $a_n = 2 + \dfrac{1}{n}$

$a_1 = 2 + \dfrac{1}{1} = 2$

$a_2 = 2 + \dfrac{1}{2} = \dfrac{3}{2}$

$a_3 = 2 + \dfrac{1}{3} = \dfrac{4}{3}$

$a_4 = 2 + \dfrac{1}{4} = \dfrac{5}{4}$

$a_5 = 2 + \dfrac{1}{5} = \dfrac{6}{5}$

In problems 11-20, find the indicated term of the sequence.

For these problems substitute the given value of n into the formula for a_n.

12. $a_n = (-1)^{n-1}$, $n = 23$

$a_{23} = (-1)^{23-1} = (-1)^{22} = 1$

14. $a_n = \dfrac{n^2 - 1}{2n + 1}$, $n = 12$

$a_{12} = \dfrac{(12)^2 - 1}{2(12) + 1} = \dfrac{143}{25}$

16. $a_n = \dfrac{2n}{n + 1}$, $n = 13$

$a_{13} = \dfrac{2(13)}{13 + 1} = \dfrac{26}{14} = \dfrac{13}{7}$

In problems 21-30, predict the sixth term of each sequence. Then write a formula for the nth term of the sequence.

22. $3, 6, 9, 12, 15, \ldots$

Here the pattern is to add 3 every time. The next term is $15 + 3 = 18$.

26. $\dfrac{4}{1}, \dfrac{5}{3}, \dfrac{6}{5}, \dfrac{7}{7}, \dfrac{8}{9}, \ldots$

Here the numerator increase by 1 every time, and the denominator increases by 2.

The next term is $\dfrac{9}{11}$.

30. $1, 2, 6, 24, 120, \ldots$

Here we multiply by 2, then 3, then 4, etc. The next term is $120 \cdot 6 = 720$.

In problems 31-40, write the first five terms of the sequence defined recursively.

In a recursively defined sequence, each term is defined using a formula that depends on one or more previous terms in the sequence. It is necessary to calculate the terms in order.

32. $a_1 = 7$, $a_{n+1} = a_n - 2$

$a_1 = 7$

$a_2 = a_1 - 2 = 7 - 2 = 5$

$a_3 = a_2 - 2 = 5 - 2 = 3$

$a_4 = a_3 - 2 = 3 - 2 = 1$

$a_5 = a_4 - 2 = 1 - 2 = -1$

36. $a_1 = 2$, $a_{n+1} = 2a_n + 3$

$a_1 = 2$

$a_2 = 2a_1 + 3 = 2(2) + 3 = 7$

$a_3 = 2a_2 + 3 = 2(7) + 3 = 17$

$a_4 = 2a_3 + 3 = 2(17) + 3 = 37$

$a_5 = 2a_4 + 3 = 2(37) + 3 = 77$

38. $a_1 = 5$, $a_{n+1} = a_n + n$

$a_1 = 5$

$a_2 = a_1 + 1 = 5 + 1 = 6$

$a_3 = a_2 + 2 = 6 + 2 = 8$

$a_4 = a_3 + 3 = 8 + 3 = 11$

$a_5 = a_4 + 4 = 11 + 4 = 15$

40. $a_1 = 3$, $a_2 = 5$, $a_{n+2} = a_{n+1} + 2a_n$

$a_1 = 3$

$a_2 = 5$

$a_3 = a_2 + 2a_1 = 5 + 2(3) = 11$

$a_4 = a_3 + 2a_2 = 11 + 2(5) = 21$

$a_5 = a_4 + 2a_3 = 21 + 2(11) = 45$

9.2   Exercises

In problems 1-6, a formula for the nth term of an arithmetic sequence is given. Write out the first four terms of the sequence. What is the first term a_1? What is the common difference d?

Just like the previous section, we substitute the values $n = 1, 2, 3, 4$ into the formula for a_n.

2. $a_n = 2n + 1$

   $a_1 = 2(1) + 1 = 3$

   $a_2 = 2(2) + 1 = 5$

   $a_3 = 2(3) + 1 = 7$

   $a_4 = 2(4) + 1 = 9$

4. $a_n = 2 + 3n$

   $a_1 = 2 + 3(1) = 5$

   $a_2 = 2 + 3(2) = 8$

   $a_3 = 2 + 3(3) = 11$

   $a_4 = 2 + 3(4) = 14$

In problems 7 - 16, find a general formula for the nth term of each arithmetic sequence. Then use the formula to calculate the 15th term of that sequence.

Here we will use the formula $a_n = a_1 + (n-1)d$. Remember that the answer will still contain the n variable.

8. $a_1 = 2, d = -2$

   $a_n = a_1 + (n - 1)d = 2 + (n - 1)(-2) = 2 - 2n + 2 = 4 - 2n$

   $a_{15} = 4 - 2(15) = -26$

10. $a_1 = 4, a_2 = 2$

   In order to calculate d, we will let $d = a_2 - a_1 = 2 - 4 = -2$.

   $a_n = a_1 + (n - 1)d = 4 + (n - 1)(-2) = 4 - 2n + 2 = 6 - 2n$

   $a_{15} = 6 - 2(15) = -24$

14. $a_2 = 15,\ a_6 = 3$

Here we have $a_2 = a_1 + (2-1)d = a_1 + d$ and $a_6 = a_1 + (6-1)d = a_1 + 5d$.

Subtracting these two yields $a_6 - a^2 = 4d$. Since $a_6 - a^2 = 3 - 15 = -12$, this

means $4d = -12$ so $d = -3$. Since $a_2 = a_1 + d$, we have $15 = a_1 - 3$ and $a_1 = 18$.

$a_n = a_1 + (n-1)d = 18 + (n-1)(-3) = 18 - 3n + 3 = 21 - 3n$.

$a_{15} = 21 - 3(15) = -24$

16. $a_1 = 3,\ a_{n+1} = a_n + 3$

Since $d = a_{n+1} - a_n$ and $a_{n+1} - a_n = 3$, we have $d = 3$.

$a_n = a_1 + (n-1)d = 3 + (n-1)(3) = 3 + 3n - 3 = 3n$,

$a_{15} = 3(15) = 45$

In problems 17-22, a formula for the nth term of a geometric sequence is given. Write out the first four terms of the sequence. What is the first term a_1? What is the common ratio r?

For these problems we will substitute the values $n = 1,\ 2,\ 3,\ 4$ into the formula for a_n.

18. $a_n = 4^n$

$a_1 = 4^1 = 4$

$a_2 = 4^2 = 16$

$a_3 = 4^3 = 64$

$a_4 = 4^4 = 256$

22. $a_n = 3\left(\dfrac{2}{3}\right)^n$

$a_1 = 3\left(\dfrac{2}{3}\right)^1 = 2$

$a_2 = 3\left(\dfrac{2}{3}\right)^2 = \dfrac{4}{3}$

$a_3 = 3\left(\dfrac{2}{3}\right)^3 = \dfrac{8}{9}$

$a_4 = 3\left(\dfrac{2}{3}\right)^4 = \dfrac{16}{27}$

In problems 23-30, find a general formula for the nth term of each geometric sequence. Then use the formula to predict the 8th term of that sequence.

In these problems we will use the formula $a_n = a_1 r^{n-1}$. Sometimes we will need to determine the values of a_1 and r from the information that is given.

24. $a_1 = \dfrac{2}{3}$, $r = 3$

$$a_n = a_1 r^{n-1} = \left(\dfrac{2}{3}\right) 3^{n-1} = 2 \cdot 3^{n-2}$$

$$a_8 = 2 \cdot 3^{8-2} = 2 \cdot 3^6 = 1458$$

26. $a_1 = 3$, $a_2 = -6$

We have $r = \dfrac{a_2}{a_1} = -\dfrac{6}{3} = -2.$

$$a_n = a_1 r^{n-1} = 3 \cdot (-2)^{n-1}$$

$$a_8 = 3 \cdot (-2)^{8-1} = 3 \cdot (-2)^7 = -384$$

30. $a_3 = 2$, $a_6 = -16$

We have $a_6 = a_1 r^{6-1} = a_1 r^5$ and $a_3 = a_1 r^{3-1} = a_1 r^2$. Dividing these equations, we get

$\dfrac{a_6}{a_3} = \dfrac{a_1 r^5}{a_1 r^2} = r^3$. Since $\dfrac{a_6}{a_3} = \dfrac{-16}{2} = -8$, we get $r^3 = -8$ so $r = -2$. Since $a_3 = a_1 r^2$,

we have $2 = a_1(-2)^2 = 4a_1$ so $a_1 = \dfrac{1}{2}.$

$$a_n = a_1 r^{n-1} = \left(\dfrac{1}{2}\right)(-2)^{n-1}.$$

$$a_8 = \left(\dfrac{1}{2}\right)(-2)^{8-1} = -64$$

9.3   Exercises

In problems 1-10, find the common difference and sum of each arithmetic series.

We will use the formula $S_n = \dfrac{n}{2}(a_1 + a_n)$. a_1 is the first term, a_n is the last term, and n is the number of terms.

2. $2 + 4 + 6 + \ldots + 20$ (10 terms)

$a_1 = 2$, $a_{10} = 20$, and $n = 10$:

$$S_{10} = \frac{n}{2}(a_1 + a_n) = \frac{10}{2}(2 + 20) = 110$$

6. $2 + 5 + 8 + \ldots + 299$ (100 terms)

$a_1 = 2$, $a_{100} = 299$, and $n = 100$:

$$S_{100} = \frac{n}{2}(a_1 + a_n) = \frac{100}{2}(2 + 299) = 15,050$$

8. $2 + 4 + 6 + \ldots + 1000$ (500 terms)

$a_1 = 2$, $a_{500} = 1000$, and $n = 500$:

$$S_{500} = \frac{n}{2}(a_1 + a_n) = \frac{500}{2}(2 + 1000) = 250,500$$

In problems 11-20, calculate the sum of each arithmetic series based upon the information given.

We have two formulas for the sum of the first n terms of an arithmetic series:

$$S_n = a_1 \cdot n + \frac{nd(n-1)}{2} \qquad\qquad S_n = \frac{n}{2}(a_1 + a_n)$$

12. $a_1 = 4$, $a_{10} = 44$, $n = 10$

We will use the second formula.

$$S_{10} = \frac{n}{2}(a_1 + a_n) = \frac{10}{2}(4 + 44) = 5(48) = 240$$

16. $a_1 = 3$, $d = 3$, $n = 20$

We will use the first formula.

$$S_{20} = a_1 \cdot n + \frac{nd(n-1)}{2} = 3(20) + \frac{20(3)(20-1)}{2} = 60 + 570 = 630$$

20. $a_1 = 2$, $a_2 = 2$, $n = 25$

We have $d = a_2 - a_1 = 2 - 2 = 0$. We will use the first formula.

$$S_{25} = a_1 \cdot n + \frac{nd(n-1)}{2} = 2(25) + \frac{25(0)(25-1)}{2} = 50$$

In problems 21-30, find the common ratio and sum of each geometric series.

For these problems we will use the formula

$$S_n = \frac{a_1(r^n - 1)}{r - 1}$$

22. $1 + 2 + 4 + \ldots$ (10 terms)

We have $r = \frac{a_2}{a_1} = \frac{2}{1} = 2$ and $n = 10$:

$$S_{10} = \frac{a_1(r^n - 1)}{r - 1} = \frac{1(2^{10} - 1)}{2 - 1} = 1023$$

26. $\frac{1}{4} + \frac{1}{2} + 1 + \ldots$ (10 terms)

We have $r = \frac{a_2}{a_1} = \frac{1/2}{1/4} = 2$ and $n = 10$:

$$S_{10} = \frac{a_1(r^n - 1)}{r - 1} = \frac{\frac{1}{4}(2^{10} - 1)}{2 - 1} = \frac{1023}{4}$$

30. $\frac{1}{2} + 2 + 8 + \ldots$ (6 terms)

We have $r = \frac{a_3}{a_2} = \frac{8}{2} = 4$, $a_1 = \frac{1}{2}$, and $n = 6$:

$$S_6 = \frac{a_1(r^n - 1)}{r - 1} = \frac{\frac{1}{2}(4^6 - 1)}{4 - 1} = \frac{1365}{2}$$

In problems 31-40, calculate the sum of each geometric series based upon the information given.

For these problems we will use the formula

$$S_n = \frac{a_1(r^n - 1)}{r - 1}$$

We may first need to determine the value of r.

32. $a_1 = \dfrac{1}{4}$, $a_{10} = 128$, $n = 10$

We have $a_{10} = a_1 r^{10-1} = a_1 r^9$, so $128 = \left(\dfrac{1}{4}\right) r^9$. This gives $r^9 = 512$, so $r = 2$.

$$S_{10} = \frac{\frac{1}{4}(2^{10} - 1)}{2 - 1} = \frac{1023}{4}$$

36. $a_1 = \dfrac{1}{9}$, $r = 3$, $n = 9$

$$S_9 = \frac{a_1(r^n - 1)}{r - 1} = \frac{\frac{1}{9}(3^9 - 1)}{3 - 1} = \frac{9841}{9}$$

38. $a_1 = 3$, $r = -2$, $n = 8$

$$S_8 = \frac{a_1(r^n - 1)}{r - 1} = \frac{2((-2)^8 - 1)}{-2 - 1} = -170$$

40. $a_1 = 1$, $a_2 = \dfrac{1}{2}$, $n = 10$

Here we have $r = \dfrac{a_2}{a_1} = \dfrac{1/2}{1} = \dfrac{1}{2}$

$$S_{10} = \frac{a_1(r^n - 1)}{r - 1} = \frac{1\left(\left(\frac{1}{2}\right)^{10} - 1\right)}{\frac{1}{2} - 1} = \frac{1023}{2048}$$

9.4   Exercises

In problems 1-10, find the common ratio and sum of each infinite geometric series.

For an infinite geometric series

$$\sum_{n=0}^{\infty} a_1 r^n = a_1 + a_1 r + a_1 r^2 + a_1 r^3 + \dots,$$

the series converges to $S = \dfrac{a_1}{1 - r}$ as long as $|r| < 1$. If $|r| \geq 1$, then the series diverges.

2. $1 + \dfrac{1}{2} + \dfrac{1}{4} + \dots$

We have $a_1 = 1$ and $r = \dfrac{a_2}{a_1} = \dfrac{1/2}{1} = \dfrac{1}{2}$. Since $|r| < 1$, this series converges.

$$S = \frac{a_1}{1 - r} = \frac{1}{1 - \frac{1}{2}} = 2$$

6. $9 + 3 + 1 + \ldots$

We have $a_1 = 9$ and $r = \dfrac{a_2}{a_1} = \dfrac{3}{9} = \dfrac{1}{3}$. Again $|r| < 1$ so the series converges.

$$S = \frac{a_1}{1 - r} = \frac{9}{1 - \frac{1}{3}} = \frac{9}{\frac{2}{3}} = \frac{27}{2}$$

8. $5 - 1 + \dfrac{1}{5} - \dfrac{1}{25} + \ldots$

We have $a_1 = 5$ and $r = \dfrac{a_2}{a_1} = -\dfrac{1}{5}$. This series converges.

$$S = \frac{a_1}{1 - r} = \frac{5}{1 - \left(-\frac{1}{5}\right)} = \frac{5}{\frac{6}{5}} = frac256$$

In problems 11-20, find the sum of each infinite geometric series.

12. $a_1 = 3$, $r = .1$

Since $|r| < 1$, this series converges. $S = \dfrac{a_1}{1 - r} = \dfrac{3}{1 - .1} = \dfrac{3}{.9} = \dfrac{10}{3}$

16. $a_1 = 5$, $r = \dfrac{2}{3}$

Since $|r| < 1$, this series converges $S = \dfrac{a_1}{1 - r} = \dfrac{5}{1 - \frac{2}{3}} = \dfrac{5}{\frac{1}{3}} = 15$

20. $a_1 = 6$, $a_2 = -2$

We have $r = \dfrac{-2}{6} = -\dfrac{1}{3}$. Again, $|r| < 1$ so the series converges.

$$S = \frac{a_1}{1 - r} = \frac{6}{1 - \left(-\frac{1}{3}\right)} = \frac{6}{1 + \frac{1}{3}} = \frac{6}{\frac{4}{3}} = \frac{9}{2}$$

22. **Grains of Wheat** In an old story, a peasant who saves the life of a king was offered a just reward. The peasant replies: I would like some wheat. The king agrees, and asks: How much? The peasant replies: Upon a chessboard, place a single grain on the first square. Then on the second square place 2 grains. On the third square place 4, and on each successive square place twice the amount in the previous square. The king agrees, thinking he got a bargain. How many grains will be required to cover all 64 squares on the chessboard?

The number of grains on each square forms a geometric sequence with $a_1 = 1$, $a_2 = 2$, $a_3 = 4$, and so on. Therefore $r = \dfrac{a_2}{a_1} = 2$ and $n = 64$. Using the formula $S_n = \dfrac{a_1(r^n - 1)}{r - 1}$, we get $S_{64} = \dfrac{1(2^{64} - 1)}{2 - 1} = 2^{64} - 1 \approx 1.844 \times 10^{19}$. This amount of wheat would bankrupt any country.

9.5 Exercises

Calculate each sum.

For these problems we will use the following formulas:

$$\sum_{i=1}^{n} 1 = n \qquad\qquad \sum_{i=1}^{n} i = \frac{n(n + 1)}{2}$$

$$\sum_{i=1}^{n} i^2 = \frac{n(n + 1)(2n + 1)}{6} \qquad\qquad \sum_{i=1}^{n} i^3 = \frac{n^2(n + 1)^2}{4}$$

2. $\displaystyle\sum_{k=1}^{10} 3k - 2$

$$
\begin{aligned}
\sum_{k=1}^{10} 3k - 2 &= \sum_{k=1}^{10} 3k - \sum_{k=1}^{10} 2 \\
&= 3\sum_{k=1}^{10} k - 2\sum_{k=1}^{10} 1 \\
&= 3\left(\frac{10(10 + 1)}{2}\right) - 2(10) \\
&= 3(5)(11) - 20 \\
&= 145
\end{aligned}
$$

4. $\displaystyle\sum_{k=1}^{12} k^2 + k$

$$
\begin{aligned}
\sum_{k=1}^{12} k^2 + k &= \sum_{k=1}^{12} k^2 + \sum_{k=1}^{12} k \\
&= \frac{12(12 + 1)(2(12) + 1)}{6} + \frac{12(12 + 1)}{2} \\
&= 2(13)(25) + 6(13) \\
&= 728
\end{aligned}
$$

8. $\displaystyle\sum_{k=1}^{15} 2k^3 - 3k$

$$
\begin{aligned}
\sum_{k=1}^{15} 2k^3 - 3k &= \sum_{k=1}^{15} 2k^3 - \sum_{k=1}^{15} 3k \\
&= 2\sum_{k=1}^{15} k^3 - 3\sum_{k=1}^{15} k \\
&= 2\left(\frac{15^2(15+1)^2}{4}\right) - 3\left(\frac{15(15+1)}{2}\right) \\
&= 2\left(\frac{225(256)}{4}\right) - 3\left(\frac{15(16)}{2}\right) \\
&= 28,800 - 360 \\
&= 28,440
\end{aligned}
$$

Calculate each expression using the binomial theorem.

The binomial theorem states that if n is a positive integer, then

$$
(a+b)^n = \sum_{k=0}^{n} \binom{n}{k} a^{n-k}b^k
$$

12. $(x+1)^4$

For this problem, we will use $a = x$ and $b = 1$ in the binomial theorem formula:

$$
\begin{aligned}
(x+1)^4 &= \sum_{k=0}^{4} \binom{4}{k} x^{4-k}(1)^k \\
&= \frac{4!}{0!(4-0)!}x^{4-0} + \frac{4!}{1!(4-1)!}x^{4-1} + \frac{4!}{2!(4-2)!}x^{4-2} \\
&\quad + \frac{4!}{3!(4-3)!}x^{4-3} + \frac{4!}{4!(4-4)!}x^{4-4} \\
&= \frac{24}{1\cdot 24}x^4 + \frac{24}{1\cdot 6}x^3 + \frac{24}{2\cdot 2}x^2 + \frac{24}{6\cdot 1}x + \frac{24}{24\cdot 1} \\
&= x^4 + 4x^3 + 6x^2 + 4x + 1
\end{aligned}
$$

14. $(x-3)^3$

For this problem, we will use $a = x$ and $b = -3$ in the binomial theorem formula:

$$
\begin{aligned}
(x-3)^3 &= \sum_{k=0}^{3} \binom{3}{k} x^{3-k}(-3)^k \\[2mm]
&= \frac{3!}{0!(3-0)!}x^{3-0}(-3)^0 + \frac{3!}{1!(3-1)!}x^{3-1}(-3)^1 + \frac{3!}{2!(3-2)!}x^{3-2}(-3)^2 \\[2mm]
&\quad + \frac{3!}{3!(3-3)!}x^{3-3}(-3)^{3-0} \\[2mm]
&= \frac{6}{1\cdot 6}x^3 + \frac{6}{1\cdot 2}x^2(-3) + \frac{6}{2\cdot 1}x^1(-3)^2 + \frac{6}{6\cdot 1}x^0(-3)^3 \\[2mm]
&= x^3 - 3(3)x^2 + 3(9)x - 27 \\[2mm]
&= x^3 - 9x^2 + 27x - 27
\end{aligned}
$$

18. $(3x-2y)^4$

For this problem, we will use $a = 3x$ and $b = -2y$ in the binomial theorem formula:

$$
\begin{aligned}
(3x-2y)^4 &= \sum_{k=0}^{4} \binom{4}{k} (3x)^{4-k}(-2y)^k \\[2mm]
&= \frac{4!}{0!(4-0)!}(3x)^{4-0}(-2y)^0 + \frac{4!}{1!(4-1)!}(3x)^{4-1}(-2y)^1 + \frac{4!}{2!(4-2)!}(3x)^{4-2}(-2y)^2 \\[2mm]
&\quad + \frac{4!}{3!(4-3)!}(3x)^{4-3}(-2y)^3 + \frac{4!}{4!(4-4)!}(3x)^{4-4}(-2y)^4 \\[2mm]
&= (3x)^4 + 4(3x)^3(-2y) + 6(3x)^2(-2y)^2 + 4(3x)(-2y)^3 + (-2y)^4 \\[2mm]
&= 81x^4 + 4(27x^3)(-2y) + 6(9x^2)(4y^2) + 4(3x)(-8y^3) + 16y^4 \\[2mm]
&= 81x^4 - 216x^3 y + 216x^2 y^2 - 96xy^3 + 16y^4
\end{aligned}
$$

9.6 Exercises

For exercises 1 - 6, determine the statements P_1 and P_{n+1} for the given statement P_n.

In these problems, we substitute either 1 or $n+1$ into each of the given statements.

2. $P_n : S_n = 3n^2 - 4n$

$P_1 : S_1 = 3(1^2) - 4(1) = -1$

$P_{n+1} : S_{n+1} = 3(n+1)^2 - 4(n+1)$

6. $P_n : S_n = 3 + 9 + 27 + \ldots + 3^n$

$\quad P_1 : S_1 = 3$

$\quad P_{n+1} : S_{n+1} = 3 + 9 + 27 + \ldots + 3^n + 3^{n+1}$

For exercises 7 - 16, use mathematical induction to prove each formula for $n > 0$.

The principle of mathematical induction is used to prove certain types of statements. We start with the case $n = 1$ and verify it is true. Then we assume that P_n is true for all values up to n, and use it to prove that P_{n+1} is true as well. This is enough to prove that P_n is true for all positive integers n.

8. $3 + 9 + 15 + \ldots + (6n - 3) = 3n^2$

This statement is true for $n = 1$ because then we are only adding one term. This gives $3 = 3(1^2) = 3$, which is true.

Now assume that the statement is true for $n > 1$. We need to prove it is also true for $n + 1$. So we are calculating $3 + 9 + 15 + \ldots + (6n - 3) + (6(n + 1) - 3)$. We assume that $3 + 9 + 15 + \ldots + (6n - 3) = 3n^2$. Then the new sum is

$$
\begin{aligned}
3 + 9 + 15 + \ldots + (6n - 3) + (6(n + 1) - 3) &= [3 + 9 + 15 + \ldots + (6n - 3)] + (6n + 6 - 3) \\
&= 3n^2 + (6n + 6 - 3) \\
&= 3n^2 + 6n + 3 \\
&= 3(n^2 + 2n + 1) \\
&= 3(n + 1)^2
\end{aligned}
$$

This is the same formula as before, with n replaced by $n + 1$.

10. $2(1 + 3 + 3^2 + \ldots + 3^{n-1}) = 3^n - 1$

For the case $n = 1$, the left hand side of this equation reduces to only one term: $2(1) = 2$. The right hand side is $3^1 - 1 = 2$, which is the same.

Now assume the statement is true for $n > 1$. So we have $2(1 + 3 + 3^2 + \ldots + 3^{n-1}) = 3^n - 1$. We need to add a term to the left hand side:

$$
\begin{aligned}
2(1 + 3 + 3^2 + \ldots + 3^{n-1} + 3^{(n+1)-1}) &= 2(1 + 3 + 3^2 + \ldots + 3^{n-1}) + 2(3^n) \\
&= 3^n - 1 + 2 \cdot 3^n \\
&= 3 \cdot 3^n - 1 \\
&= 3^{n+1} - 1
\end{aligned}
$$

The final answer is the same as $3^n - 1$ with n being replaced by $n+1$. Therefore the statement is true for all $n > 0$.

12. $\displaystyle\sum_{i=1}^{n} i^3 = \frac{n^2(n+1)^2}{4}$

First we need to verify the formula for $n = 1$. The equation becomes $\sum_{i=1}^{1} i^3 = \frac{1^2(1+1)^2}{4} = 1$. This is true.

Next assume the formula is true for $n > 1$, that is, $\displaystyle\sum_{i=1}^{n} i^3 = \frac{n^2(n+1)^2}{4}$. Replace n with $n+1$ on the left hand side. This becomes $\displaystyle\sum_{i=1}^{n+1} i^3 = \sum_{i=1}^{n} i^3 + (n+1)^3 = \frac{n^2(n+1)^2}{4} + (n+1)^3$.

We need to simplify this:

$$
\begin{aligned}
\frac{n^2(n+1)^2}{4} + (n+1)^3 &= \frac{n^2(n^2+2n+1)}{4} + \frac{4(n+1)^3}{4} \\
&= \frac{n^4 + 2n^3 + n^2}{4} + \frac{4(n^3 + 3n^2 + 3n + 1)}{4} \\
&= \frac{n^4 + 2n^3 + n^2 + 4n^3 + 12n^2 + 12n + 4}{4} \\
&= \frac{n^4 + 6n^3 + 13n^2 + 12n + 4}{4} \\
&= \frac{(n+1)^2(n+2)^2}{4}
\end{aligned}
$$

This last expression is the same as the original right hand side of the formula, with n replaced by $n+1$.

For exercises 17 - 20, use mathematical induction to prove each inequality for $n > 0$.

18. $4^n > n2^n$

First we verify the inequality for $n = 1$. This gives $4^1 > 1(2^1)$, which is $4 > 2$, which is true.

Now assume it is true for $n > 1$. Then $4^n > n2^n$. We need to prove that $4^{n+1} > (n+1)2^{n+1}$. However, $4^{n+1} = 4 \cdot 4^n > 4n2^n = 2^2 n2^n = 2n2^{n+1} > (n+1)2^{n+1}$. The last inequality is true because $2n > n+1$ if $n > 1$. This proves the statement.

20. $\left(\dfrac{1}{x}\right)^{n+1} < \left(\dfrac{1}{x}\right)^{n}, x > 1$

First we need to prove this statement for $n = 1$. The statement becomes $\left(\dfrac{1}{x}\right)^{1+1} < \left(\dfrac{1}{x}\right)^{1}$, or $\dfrac{1}{x^2} < \dfrac{1}{x}$. If $x > 1$ then we can multiply both sides of the inequality by x^2, getting $1 < x$, which is true.

Now assume it is true for $n > 1$. Then we have $\left(\dfrac{1}{x}\right)^{n+1} < \left(\dfrac{1}{x}\right)^{n}$. We need to prove the statement is true for $n + 1$. Then the inequality becomes $\left(\dfrac{1}{x}\right)^{(n+1)+1} < \left(\dfrac{1}{x}\right)^{n+1}$. If we multiply both sides by x^{n+2} then the inequality becomes $1 < x$, which is true. This proves the original statement.